The Spacetime Origin Of the Universe

by
Vladimir B. Ginzburg

Edited by Ellen Orner
Cover by Eugene B. Ginzburg

First Edition

Helicola Press
Division of IRMC, Inc.
612 Driftwood Drive
Pittsburgh, Pennsylvania, 15238

The Spacetime Origin of the Universe

The First Edition

Published by
Helicola Press
Division of IRMC, Inc.
612 Driftwood Drive
Pittsburgh, Pennsylvania
USA
www.helicola@aol.com

Printed in the United States of America

ISBN: 978-0-578-12546-6

Current printing (last digit)
10 9 8 7 6 5 4 3 2 1

Front book cover image credit:
NASA/JPL-Caltech/ESA/Harvard-Smithsonian CfA

My special thanks go to the members of my family: to my gorgeous, friendly and energetic grandson Alex, to my beautiful granddaughter Alexandra, who since the age of 10 became the youngest scientifically-minded person capable of understanding my theory; to my daughter Ellen, with whom I brainstormed some ideas related to the vortex theory, and who edited my three previous books on this subject and also Chapter 1 of this book; to my son Gene, who provided many digital illustrations for my books, and also designed the covers of all my books including this one; to my brother Paul who patiently followed my research in this field and provided me with his valuable comments and corrections, and, finally, to my wife Tanya, whose moral support, advice, and assistance in editing this book were invaluable.

Contents

THE ESSENCE

OF THE UNIVERSAL SPACETIME THEORY

This book describes the basic concept of the Universal Spacetime Theory (UST). During its development I was stimulated by many scientific ideas proposed since the dawn of science described briefly in Chapter 1. In the UST, these ideas took on a completely new form, thanks to my discovery of a spiral string entity that I named *the toryx*. The toryx turned out to be a perfect candidate for a prime element of nature. This, however, was only the tip of the iceberg. Besides yielding the unified laws applicable to both micro- and macro-worlds, the toryx helped me to find the answer to the most puzzling question of science:

What is the origin of the universe?

The toryx is the smallest entity that retains three fundamental properties of the universe: spacetime, spirality and polarization.

Spacetime - Welcome to the spacetime-based universe in which the spacetime parameters are indispensible attributes of all its components. Unlike many physical parameters, such as mass, charge, force and energy, invented by the inhabitants of our planet, the spacetime parameters are universal and can readily be discovered by the inhabitants of any civilization of the universe.

Spirality - Welcome to the spiral universe in which spirality pervades visibly and invisibly. The spirality reveals itself in various forms, including spiral galaxies, spiral paths of celestial bodies, hurricanes, tornadoes, water vortices, and the double-helical structure of DNA.

Polarization - Welcome to the polarized universe in which oppositely-charged electrons and protons coexist inside all atoms. The universe in which there is no young without old, there is no hot without cold, there is no high without low, there is no fast without slow, there is no dark without light, there is no left without right, there is no negative without positive, there is no polarization without unification.

Discovery of the Toryx – I discovered the toryx in 1993. Convinced in the spiral character of the universe, I began to investigate whether spirals are also present in atoms. The work of the French physicist Louis De Broglie gave me a good starting point.

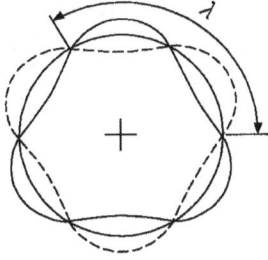

Figure 01. Standing waves fit to a circular Bohr orbit of an electron for $n = 3$. Adapted from R.A. Serway (1990).

In 1923 de Broglie made a proposition that the motion of an electron in the hydrogen atom is accompanied by a two-dimensional standing wave with the wavelength λ, which guides the electron along its path around a proton (Figure 01). So, I wanted to investigate what happens if I replace de Broglie's two-dimensional standing wave with a three-dimensional toroidal spiral shown in Figure 02.

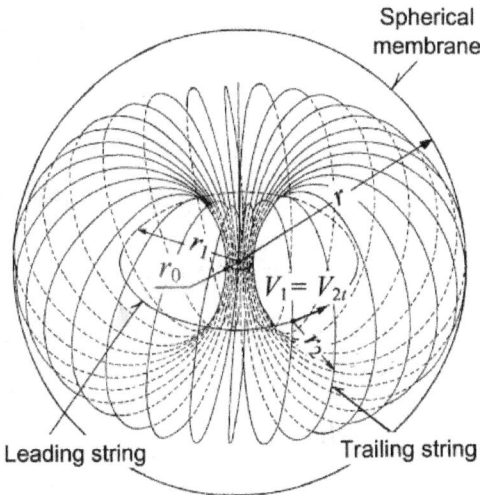

Figure 02. Toryx.

The obtained spacetime entity (toryx) consisted of three main parts:

a) Circular leading string with the radius r_1 propagating along its circular path with the velocity V_1.
b) Toroidal trailing string with the radius r_2 propagating along its toroidal spiral path with the spiral velocity V_2 and, at the same time, moving along the circular leading string with the translational velocity V_{2t} equal to V_1.
c) Spherical membrane with the radius r embracing the toryx trailing string and serving as a part of the toryx fine structure.

My next step was to establish the relationships between the toryx spacetime parameters such as radii, wavelengths, spiral lengths, steepness angles, number of windings, frequencies, periods and velocities. To accomplish this task it was necessary to limit the toryx's degrees of freedom. Otherwise the toryx would have an infinite number of possible shapes and dimensions, making the theory completely useless.

After investigating several ways of limiting the toryx's degrees of freedom, I found the best solution by using three very simple fundamental spacetime equations that I called the "toryx sacred code."

1. The length of one winding of trailing string L_2 is equal to the length of one winding of leading string L_1:

$$L_2 = L_1$$

2. The toryx eye radius r_0 is equal to a real positive constant:

$$r_0 = const.$$

3. The spiral velocity of trailing string V_2 is constant and equals to the velocity of light in vacuum c:

$$V_2 = c = const.$$

This was it. The three simplest fundamental spacetime equations of the "toryx sacred code" allowed me to derive equations establishing the relationships between all spacetime parameters of the toryx. Expecting no surprises, I began a standard routine analysis of these equations allowing me to see what happens with the toryx shape when the radius of its leading string r_1 changes from positive to negative infinity. The results of my analysis were overwhelming. As the as the radius of the toryx leading string r_1 decreased, its shape underwent through four completely unexpected transformations involving inversion (or turning inside out) of its leading string, trailing string and spherical membrane.

But this was not all I got. I discovered that these topological transformations provided the toryces with the capability to exist in polarized states and to serve as the prime elements of elementary matter particles.

Elementary matter particles – I made a major step forward after realizing that the unification of polarized toryces could produce four kinds of elementary matter particles: *ethertrons, singulatrons, electrons* and *positrons*. Figure 03 shows the formation of elementary matter particles as a function of the steepness angle of the troryx trailing string φ_2 and the relative radius of its spherical membrane b.

Ethertrons are formed by the unification of the real toryces $A^+_{m.n.q}$ and A^-_{mnq} with opposite vorticities. They are the lightest elementary matter particles serving as the principal constituents of *ether* and *quantum vacuum*.

Singulatrons are formed by the unification of the imaginary toryces $\breve{A}^+_{m.n.q}$ and $\breve{A}^-_{mnq}$ with opposite vorticities. They are the heaviest elementary matter particles serving as the principal constituents of *nucleon cores* and *singularities*.

Positrons are formed by the unification of the toryces E^+_{mnq} and $\breve{E}^+_{m.n.q}$ with positive vorticities and opposite realities. They serve as the inner shells of protons and neutrons.

Electrons are formed by the unification of the toryces $E^-_{m.n.q}$ and $\breve{E}^-_{mnq}$ with negative vorticities and opposite realities. They serve as the outer shells of neutrons and atoms.

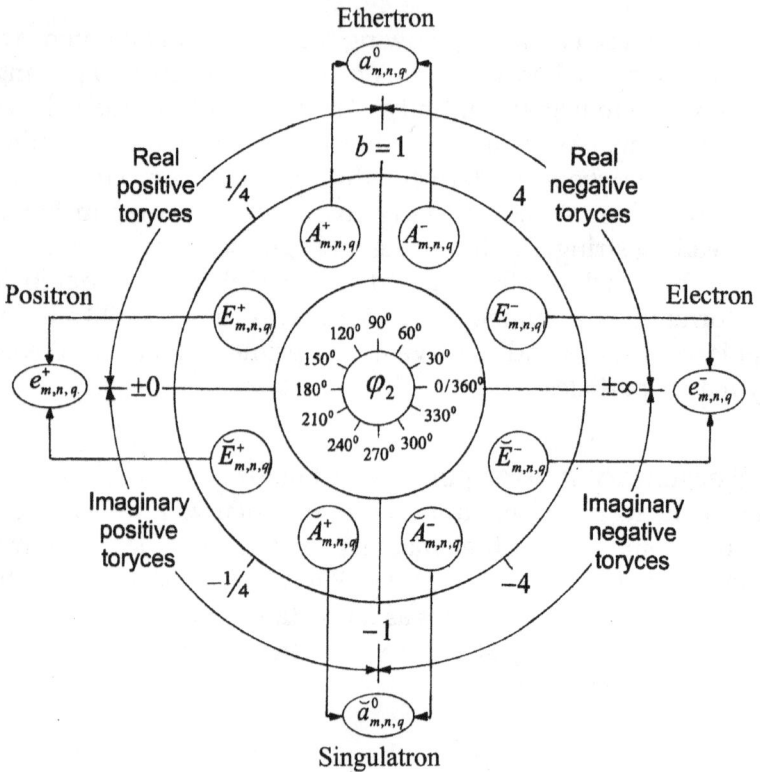

Figure 03. Formation of four elementary matter particles.

Finally, I found that at the atomic and subatomic levels physical properties of toryces can be expressed as functions of their spacetime properties. Consequently, physical constants may be thought of as merely conversion factors between man-invented physical properties of toryces and their universal spacetime parameters. I found that the toryces must exist in various quantum energy states indicated by the subscripts m, n and q in their symbols. This provides toryces with two additional capabilities:

- The capability to get excited and oscillated and, consequently, emit elementary radiation spacetimes called *helyces* forming elementary radiation particles propagating with both luminal and superluminal velocities.
- The capability to form nucleons and atoms of both ordinary and dark matter.

My most shocking discovery of all was that the elementary matter particles retained the most vital traits of living entities, the ability to sustain their existence by the absorption and release of energy.

Unified Laws of the UST – The UST yields two unified laws that govern the existence of entities made up of toryces.

Unified Law of Planetary Motion – Similarly to the law of planetary motion of planetary systems based on classical mechanics, this law establishes the relationship between the velocity of a body orbiting a central body and its distance to the central body. I found that for large distances between bodies both the unified and classical laws of planetary motion yielded the same results with a negligibly small difference. Mathematically, Kepler's Third Law of Planetary Motion is a particular case of the Unified Law of Planetary Motion.

Unified Law of Stable Polarization – According to this law, an enclosed system is stable if the sum of polarization parameters of all its components is infinitely small. Therefore, independent on amplitudes of polarization of its constituent toryces, the matter particles are stable if the sum of polarization parameters of their constituent toryces is infinitely small.

Conclusions – Based on the results of my research, I made the following conclusions regarding to the origin and existence of the universe.

- The universe has always been an assembly of polarized toroidal spiral spacetimes called *toryces* with various amplitudes of their polarization parameters. It is governed by two fundamental laws: the Unified Law of Planetary Motion and the Unified Law of Stable Polarization.
- The toryces are capable to get excited and oscillated and, consequently, emit elementary radiation spacetimes called *helyces* forming elementary radiation particles propagating with both luminal and superluminal velocities.
- The masses of the toryces are proportional to the amplitudes of their vorticities. Consequently, the toryces making up singulatrons have the greatest masses, while the toryces making up ethertrons the smallest masses.
- Similarly to the electrons, protons can also be in excited energy states. Some physicists working at the CERN Large

Hadron Collider believe that they discovered Higgs boson. According to the UST, it is merely an excited proton.

- In the most extreme case, the universe would be in the form of a *quantum vacuum* containing a singularity with an infinitely large mass occupying a tiny spot and surrounded by an infinite number of ethertrons with infinitely small masses (Figure 04).
- Associated with all bodies of macro-world are the macro-toryces based on the same fundamental space time equations as the toryces of micro-world. The toryces of the macro-world govern the interactions between all bodies of macro-world.

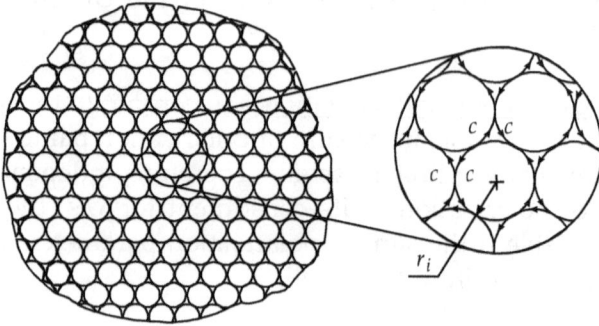

Figure 04. Quantum vacuum.

The rest is commentary, as described in this book.

ACKNOWLEDGEMENTS

In the first Chapter of this book I gave credits to many professional and amateur scientists of all times whose ideas are related to the Spiral string Theory. I am very grateful to two scientists who provided their valuable contributions to this Chapter. The American theoretical physicist Richard Gauthier outlined briefly his transluminal energy quantum (TEQ) models of photon, electron, cosmic quantum and dark matter. The American-Australian molecular biologist Horace R. Drew, the co-author of groundbreaking book *Understanding DNA – The Molecule and How It Works*, wrote an article on the helical theory of DNA.

I am also grateful to several scientists for making their valuable and stimulating comments on my novel ideas. Among them are: Professor Carlo Rovelli (Department of Physics and Astronomy of the University of Pittsburgh), Professor Gregory M. Townsend (Department of Physics of the University of Akron), Professor Clifford Taubes (Department of Mathematics, Harvard University), who made encouraging comments on my idea of multiple-level Universe.

Several American physicists made valuable comments on my model. Among them are: Professor Rudolph Hwa (Department of Physics, University of Oregon), Dr. Blair M. Smith (Innovative Nuclear Space Power and Propulsion Institute, University of Florida), Professor Edward F. Redish (Department of Physics, University of Maryland), Professor David F. Measday (Department of Physics and Astronomy, University of British Columbia, Canada), Dr. Eric Carlson (Department of Physics, Wake Forest University), and Professor Warren Siegel (Department of Physics and Astronomy, Stony Brook University).

Thanks to the recommendations made by Dr. Akhlesh Lakhtakia (Department of Engineering and Mechanics, the Pennsylvania State University), I was able to publish the first three papers describing my version of the vortex theory. Several my papers on this subject were published with the assistance provided by Dr. Harold Fox (Editor of *The Journal of New Energy*), Yasuhiko Genku Kimura (Publisher and Editor-in-Chief of *VIA, The Journal of Integral Thinking for Visionary Action*). I appreciate very much my interesting discussions with late Marvin Solit, Director of Foundations for New Directions, and his ideas about possible space orientations of toryces.

I am very grateful to Dr. Cynthia Kolb Whitney (Editor of the Journal *Galilean Electrodynamics*) for reading my book *Prime Elements of Ordinary Matter, Dark Matter & Dark Energy* and giving me very valuable suggestions. Mr. Nobuyuki Kanai from Osaka, Japan followed my research on spirals for the last several years and provided me with many useful corrections of my writings. I will always remember very stimulating discussions with the late Distinguished Professor Eli Gorelik of the University of Pittsburgh, Pennsylvania about a crystal structure of a nucleon core.

Let me finish the introduction to this book with the words of one great visionary of the 20th century.

That the man calls matter, or substance, has no existence whatsoever. So-called matter is but waves of the motion of light, electrically divided into opposed pairs, then electrically conditioned and patterned into what we call various substances of matter. Briefly put, matter is the motion of light, and motion is not substance. It only appears to be. Take motion away and there would not be even the appearance of substance.

Walter Russell
The author of *The Secret of Light.*

CHAPTER 1

IDEAS RELATED TO
THE UNIVERSAL SPACETIME THEORY

As many theories of physics, the Universal Spacetime Theory (UST) is based on several earlier theories developed since the beginning of science. Although it is very difficult for me to evaluate the effect of all of them on the way of my thinking, still the vortex theory was certainly the most influential among them. Introduced at the dawn of science almost 2600 years ago, it had passed through several phases of gaining strength by absorbing the discoveries made by the Greek civilization, the Copernicus Revolution, the age of electromagnetism, the atomic age, and the information age. Each time this theory managed to engage the attention of a new generation of brilliant scientists, enchanting them by the deep physical meaning of its basic concept. Still, although they employed the latest advances in science, none of them was able to produce a mathematical tool to make the vortex theory usable in practice. Consequently, this wonderful theory repeatedly faded from view, awaiting another chance to return.

It appears that its lucky chance finally arrived by the end of the 20th century with the discovery of special spiral strings called the *toryx* and the *helyx*. I provided below brief descriptions of the indirect contributions made by both professional and amateur scientists to the development of the *Universal Spacetime Theory* (UST).

1.1 The Age of the Ancient Greek Civilization

Anaximander of Miletus (611-547 BC) – The Greek natural philosopher Anaximander of Miletus proposed the idea of vortex (δίνη or δῖνος) that closely resembled the mythological "cosmic whirl." At any given time, he professed, there exists an infinite number of worlds that have been separated from the infinite, τόάπειρον, which is eternal and ageless, and also the source and reservoir of all things. After our world was created, a rotary motion in a vortex caused the heavy materials to concentrate at the center, while masses of fire surrounded by a mass of air went to the periphery and later formed the heavenly bodies. The created worlds, however, do not exist forever, and, when perished, they are absorbed back into the infinite.

Anaximander logically argued that the source from which the world begins is not identical with any of the ordinary stuff known to us; it must have the capability of giving rise to the wide variety of things of contrary qualities. According to Anaximander, the things appeared from the boundless source, or *apeiron*, that functions as a storehouse of the world's qualities and provides a mechanism for releasing and absorbing these qualities. This is a never-ending cyclic process. During each cycle, the world's qualities are separated from the apeiron by the rotary motion of a vortex, and as soon as their contrary qualities become manifest, they, in turn, become reabsorbed into the apeiron.

Pythagoras of Samos (c.560-480 BC) – In ancient Greece, there was probably no other philosopher whose name was associated with so much praise and controversy. He founded a religious and philosophical Secret Society devoted to exploring the mysteries of numbers. Through this society, Pythagoras promoted a "rounded" education, introducing *the fourfold way* of study that included geometry, arithmetic, music and astronomy. One of his greatest achievements was the number theory. He and his followers, the Pythagoreans, looked for a direct connection between the whole numbers and geometrical shapes.

Pythagoras came to the conclusion that the numbers themselves expressed not only harmonies and shapes, but the essences of the things having those harmonies and shapes. Moreover, the numbers had to be the generators of everything else. Pythagorean music theory was based upon observations of a monochord, one of the most commonly used instruments of those times. Pythagoras saw a great similarity between the ratios in geometrical figures and the ratios that divided the strings of the monochord, so making music was merely another way of using proportions between numbers. He eventually applied this idea to everything in nature.

The concept of harmony led Pythagoras to a more detailed examination of the relationships between the sides of right triangles that he summarized in his famous Pythagorean Theorem:

In any right triangle, the sum of the squares of the two sides A and B is equal to the square of the hypotenuse C, so:

$$A^2 + B^2 = C^2$$

Although the Egyptians and the Babylonians discovered many right triangles to which this rule was applied, Pythagoras stated that the rule, applicable to the right triangles known before his time, was also true for "any" right triangle. By making this theoretical generalization of practical geometry, Pythagoras made another important step towards the conversion of practical geometry into science. Ironically, Pythagoras's obsession with the idea of harmony forced him to apply his own theorem only to rational numbers. After finding that some applications of his theorem produced irrational numbers, such as the square root of 2, he ruthlessly tried to suppress this discovery, an unwise decision that contributed to his tragic death and the demise of his society.

Anaxagoras of Glazomenae (500-c.428 BC) - In his book *The Nature of Things*, the Greek philosopher Anaxagoras introduced his version of the creation of the universe by the vortex motion. Like Anaximander, made an ambitious attempt to employ vortex motion to explain the creation of the universe without reference to supernatural forces. But, unlike Anaximander, who did not explain the cause of the vortex motion, he proposed that the primordial soup was set in rotation by an all-pervading intelligence, or Mind.

To explain the multiplicity of things in the universe as well physical and biological change in those things, he proposed that the primordial soup was made of an infinite variety of substances. Consequently, as the different things separated from the primordial soup, they acquired various properties. The main argument of Anaxagoras was that there is something in everything. For instance, the water is perceived to be water because most of its parts are water, but in addition to water it contains a part of every other thing in the universe. According to Anaxagoras, besides the creation of the things with various properties, there was another important role of the vortex motion of the primordial soup. The vortex motion supplied heat through friction that ignited the sun and the stars. His system enabled Anaxagoras and his followers to describe all existing objects.

Empedocles of Acragas (c.490-c.430 BC) - the Greek philosopher Empedocles created vortices so that anyone could see what they looked like and understand how they worked. To explain the formation of celestial bodies by vortex motion, Empedocles dropped tea leaves in the center of the urn (Fig. 1.1.1), while he continued to stir the liquid. It was clearly visible that the leaves migrated towards the center and accumulated there as long as they were heavy enough to withstand the updraft. This effect came to be known to us as the "tea cup phenomenon." Empedocles thought that the Earth hovered in the updraft of the primordial vortex.

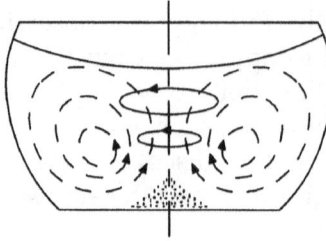

Figure 1.1.1. An illustration of the tea cup phenomenon.
Adapted from H.J. Lugt (1995).

Democritus of Abdera (c.470-c.400 BC) - The Greek philosopher Democritus of Abdera is well-known for his pioneering work on atomic theory, but he also made an important contribution to vortex theory. According to his vortex theory, the primordial soup was made of solid bodily atoms and empty space. The atoms came in various shapes and sizes, and their small dimensions made them invisible. The atoms were indestructible, homogeneous in substance, contained no void and no interstices. They were in perpetual motion in the infinitely extended void, probably moving equally in all directions. When a group of atoms became isolated, a vortex was produced by random collisions of atoms. Consequently, atoms of irregular shape became entangled and moved towards the center. Democritus considered the vortex motion so fundamental that he interpreted it simply as a general law of nature. This was probably the first attempt in the history of science to formulate a "unified theory of physics."

Archimedes of Syracuse (c.287-212 BC) – The Greek mathematician, physicist and inventor Archimedes of Syracuse made an extraordinary contribution to the development of a mathematical concept of spirals and its practical application. He discovered one of the simplest forms of spirals; it is still associated with his name.

In his work *On the Measurement of the Circle* he used the so-called *method of exhaustion* to calculate a more accurate value of π (*pi*). He put regular polygons inside and outside of a circle (Fig. 1.1.2). He then figured out their perimeters. The perimeter of the circle itself had to lie somewhere in between. After increasing the number of sides of the polygon to 96, Archimedes was able to calculate a very accurate value of π (*pi*) for his time, approximately between 3.141 and 3.143.

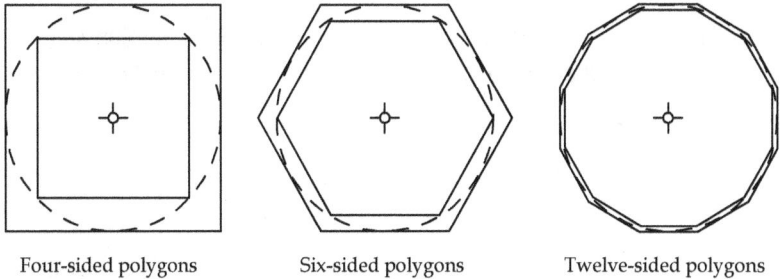

Four-sided polygons	Six-sided polygons	Twelve-sided polygons

Figure 1.1.2. Archimedes' method for determining the value for π (pi).

Archimedes realized that the most accurate result would be obtained when the number of sides was infinitely large. Another trend became obvious to him. As he continued to increase the number of sides of the polygon, the length of each side become shorter and shorter, approaching zero. Here, Archimedes saw both zero and infinity face-to-face; he also saw their close relationship, as the appearance of one of them led to the appearance of the other.

Perhaps this encounter with zero and infinity encouraged Archimedes to devise a scheme for naming the numbers that went beyond the limits of the language of his time. A myriad was the largest number used by the ancient Greeks. It was equivalent to our ten thousand. His scheme included any number up to $10^{80,000,000,000,000,000}$ in contemporary notations. This number survived as the largest number known to humanity for more than two millennia. A new term, *googol*, was coined by an American mathematician in 1955.

1.2 The Age of Classical Mechanics

Vortex theory was revived in the beginning of 16th century. Its revival was clearly associated with the advent of the Copernican revolution.

Johannes Kepler (1571-1630) - The German astronomer and physicist Johannes Kepler is well-known for his Three Laws of Planetary Motion that he derived from the precise astronomic observations made by the Dutch astronomer Tycho Brache (1546-1601). The most relevant to our theory is the Kepler's Third Law of Planetary Motion that states:

> The cubes of the mean distances of the planets from the Sun R are proportional to the squares of their periods of revolution T, so:

$$R^3 = kT^2$$

where k is constant.

While working on the development of his laws, Kepler was preoccupied with finding natural causes for the planets' motion according to these laws. He identified three causes. The first cause, he thought, was a "magnetic force," emanated from the center of the Sun. Only many years later, well after Kepler's time, was it discovered that, unlike gravitational forces, magnetic forces acting between celestial bodies are too weak to explain celestial mechanics. The second cause was a natural desire of the planets to have a "sphere, or a layer of space within which the planet must always be found." To avoid wasting of space, he proposed that the spheres of adjacent planets must be touch with each other through the Platonic solids. Finally, to explain more accurately the physical meaning of his Third Law, Kepler imagined that the radii of the orbits of various planets must have ratios to one another, much like sound frequencies of an imaginary celestial organ that plays its eternal harmonies throughout the world.

René Descartes (1596-1650) – The French philosopher and mathematician René Descartes is mostly known for his discovery of analytical geometry. It is much less known that he also developed a general theory of the universe. According to his idea, known as the *theory of vortices*, the planets floated in an ethereal fluid composed of fine globular particles whirling in the form of vortices about the Sun. He envisioned that the stars and the circling planets were simply the visible parts of a great celestial whirlpool or vortices (Fig. 1.2.1).

He realized that every natural body in the Universe could be at rest in respect to the local matter that belonged to the vortices, and yet moving with respect to distant bodies. Therefore, the Earth did not move freely through space, but was carried around the Sun in a vortex of matter without changing its place with respect to the surrounding substance. In this way, it could be said to be "stationary." The phenomenon of whirlpools in rivers provided Descartes with an analogy to explain the mechanism of planetary motion. He envisioned the matter in the sky, where the planets were turning around the Sun, as a vortex with the Sun at its center. The parts of the vortex matter that were nearer to the Sun moved faster than those that were farther away.

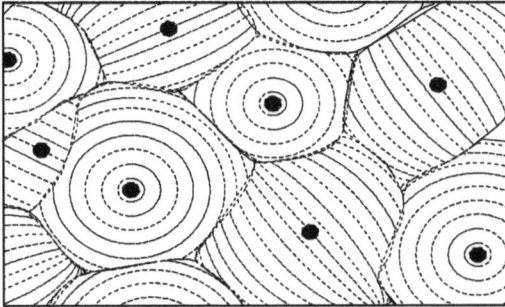

Figure 1.2.1. Congruence of celestial vortices according to Descartes. Adapted from E.J. Aiton (1972).

By using this "self-evident" analogy, Descartes was able to describe qualitatively the main observed characteristics of the planetary system, including the faster velocities of the planets located closer to the Sun and the noncircular shapes of planetary orbits. This paradigm, however, turned out to be the principal weakness in his theory. Since his fundamental cosmology was nearly all qualitative, without any quantitative and mathematical support, he came to fear that he produced nothing more than a beautiful "romance with nature."

Christiaan Huygens (1629-95) – Dutch physicist and astronomer Christiaan Huygens made several outstanding contributions to the development of science. He developed further Descartes' attempts to explain gravity by assuming that a vortex of particles of subtle matter was circling the Earth with great velocity. In his theory, Huygens replaced the cylindrically symmetrical vortices, proposed by Descartes, with a multilaterally moving vortex, in which these particles circled the Earth in all directions.

He was the first scientist who proposed an equation for the calculation of a centrifugal force. He discovered that the centrifugal force F_c, applied to a body with the mass m that orbits another body located at the distance r with the orbital velocity V, is equal to:

$$F_c = -\frac{mV^2}{r}$$

In 1679 Huygens introduced his wave theory of light. According to this theory, light propagates through ether. In Huygens' model, ether is made of small, uniform, elastic particles compressed very close together. Light propagates with very great, but finite, velocity by creating hemispherical waves around the ether particles that the light touches. Huygens' Principle stipulates that spherical waves created by single particles are too weak to transmit light, but light is transmitted when spherical waves of many particles overlap, forming a *wave front*. For his theory to work, Huygens needed ether as a medium in which the waves can propagate. He thought of ether as an "ethernal matter," which vibrates in response to a source of light, allowing the light signal to propagate.

Isaac Newton (1642-1727) – The English physicist and mathematician made a great contribution to the invention of calculus and to the development of theories of mechanics, optics and gravitation. Relevant to our story is Newton's Universal Law of Gravitation, which states:

Every particle in the universe attracts every other particle with a force F_g that is directly proportional to the product of their masses m_1 and m_2 and inversely proportional to the square of the distance r between them:

$$F_g = \frac{m_1 m_2 G}{r^2}$$

where G is Newtonian constant of gravitation.

Notably, Newton's theory of gravity offered no means of identifying any mediator of gravitational interaction. His theory assumed that gravitation acts instantaneously, regardless of distance. There was a clear distinction between the explanation of gravity offered by Newton's Universal Law of Gravitation, and that proposed by vortex theory. A dispute between Newton and his followers (Newtonians) and the followers of Descartes (Cartesians) became known as the "battle of the century." Among the most prominent Cartesians were the Dutch astronomer and physicist Christiaan Huygens and the German mathematician and philosopher Gottfried Leibniz (1646-1716).

The main thrust of vortex theory was to understand the physics of the process that made celestial bodies move along their orbits in space. From that point of view, this theory was very successful, because it explained many observed phenomena. One could readily visualize the forces acting on small debris in the vortices of a river and be easily persuaded that the vortices serve as a medium for transferring these forces. The problem, however, was that this explanation was largely qualitative. When it came to a precise comparison of its predictions against astronomical observations, vortex theory failed miserably.

The opposite was true with Newton's Universal Law of Gravitation. Its predictions worked very well, but nobody could comprehend the physical meaning of the term *attraction force*, nor explain what created this force or how it was transmitted across the vast distances of empty space from one celestial body to another. This lack of physical meaning in the theory of attraction force was of great concern to many scholars, including Newton. Therefore, for some time he couldn't decide whether or not to disregard the vortex theory.

Eventually, Newton put aside the problem of defining the physical meaning of the forces and concentrated on obtaining mathematical equations for determining the mechanics of celestial bodies in the presence of attraction forces caused by gravity and repulsion centrifugal forces caused by inertia. He looked at the circular motion of a celestial body having the mass m_2 as a state of equilibrium between two equal and opposing forces applied to this body: the centripetal gravitational force F_g generated by a central body with the mass m_1 and the centrifugal force F_c (Fig. 1.2.2). As Huygens discovered earlier, the centrifugal force F_c, applied to a body with the mass m_2 that orbits with the orbital velocity V another body located at the distance r, is equal to:

$$F_c = -\frac{m_2 V^2}{r}$$

Consequently, by equating the centrifugal force F_c and gravitational force F_g to each other ($F_c = F_g$), Newton obtain the equation expressing orbital velocity V as a function of orbital radius r.

$$V = \sqrt{\frac{m_1 G}{r}}$$

The good thing about the above equation was that it inevitably led back to Kepler's Third Law of the planetary system. This fact was used by Newton as confirmation of his theory. As we all know, his equations worked, but the price he paid, the loss of physical meaning, was too high even for his loyal supporters.

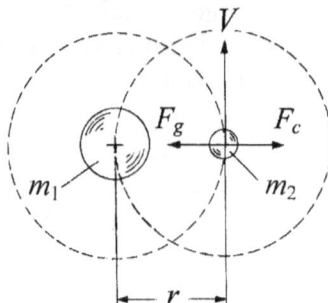

Figure 1.2.2. Celestial planetary system with two bodies.

Emanuel Swedenborg (1688-1772) – The Swedish physicist Emanuel Swedenborg made a serious attempt to challenge Newton's mechanics. He embraced this subject wholeheartedly by following his own path. Instead of applying this theory to the macro-world of celestial bodies, as Descartes, Huygens and Leibniz had done, he concentrated his efforts on applying it to the micro-world of atoms and elementary magnets. He wanted to discover the basic roots of the spiral motion in the universe. In 1734 he published a voluminous book called *Principia Rerum Naturalium*, or simply, *Principia* published. According to the concept that Swedenborg presented to the world, physical reality developed from an elementary particle that is a mathematical point. In the elementary particle, "there is an internal condition tending to a spiral motion." Through this internal condition implanted in the elementary particle, a series of material particles developed, eventually producing the Universe in its present state.

A significant part of his *Principia* was dedicated to explaining the causes and mechanisms of magnetic forces. After stating that "in every particle there is a force tending to spiral gyration," Swedenborg then clearly demonstrated that two magnets with concordant spiral gyrations would attract each other while the magnets with opposite spiral gyrations will repulse each other. In spite of the success of his book, Swedenborg nevertheless knew that he was far from reaching his ultimate goal of challenging Newton's *Principia*. He soon realized that his concept of spiral motion as the origin of matter did not capture the hearts of European scientists, who were still enchanted by the simplicity and practical validity of Newton's Laws.

Rudjier Bośković (1711-1787) - In 1758 a challenge to both vortex theory advanced by Leibniz and to Newtonian mechanics came from the Croatian physicist, astronomer and mathematician Rudjier Bośković. He formulated his principal concept in natural philosophy in his book *Philosophia Naturalis Theoria*. According to Bośković's natural philosophy, the prime elements of matter are real, homogeneous, simple, indivisible, non-extended geometrical points that possess inertia and mutual interaction.

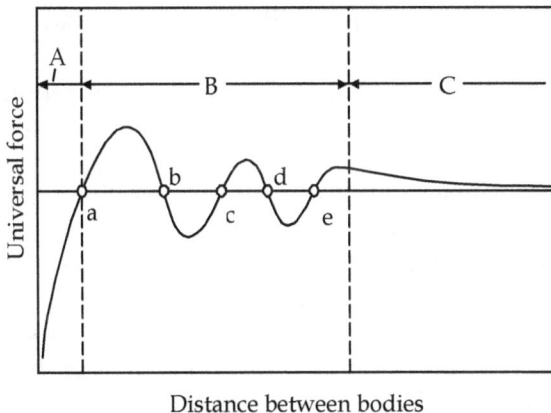

Figure 1.2.3. Graphic presentation of Bośković's Universal Force Law.

Based on that premise, Bośković challenged the validity of Newton's Universal Law of Gravitation for the case when the distances between two elements were so small that they would penetrate each other. This, however, contradicted the main premise of indivisibility of prime elements. Bośković resolved the contradiction (Fig. 1.2.3) by assuming that Newton's law remained valid only when the distance between the bodies is very large (zone C). As this distance decreases (zone B), the force ceases to follow

Newton's law and becomes alternatively either attractive or repulsive, depending upon the distance by which the bodies are separated. Then at even shorter distance (zone A), the force becomes purely repulsive. As the distance diminishes to zero, repulsion grows infinite, thus preventing direct contact between particles. Bošković called the proposed relationship between the forces acting on two particles the *Universal Force Law*.

Bošković further speculated that the points of equilibrium between attraction and repulsion forces *a, b, c, d, e*... played a very important role. These points, which he called "boundaries," correspond to either stable or unstable equilibrium between the bodies. By evaluating the behavior of these boundaries and the area of the force-versus-distance curve contained between these boundaries, Bošković was able to explain various physical and chemical properties of matter, including emission of light. This was certainly the first attempt to explain such a great variety of natural phenomena by using a *unified theory*.

1.3 The Age of Electromagnetism

Reincarnation of the vortex theory in the 18th century was triggered by the advent of the age of electricity and magnetism.

Benjamin Franklin (1706-1790) – Benjamin Franklin became the first American to win an international reputation in pure science and the first man of science to gain fame for work done wholly in electricity. His most important achievements included the formulation of the theory of general electrical action that explained the process of production and transfer of electrostatic charges. It had long been common knowledge that by conducting rather simple experiments it is possible to demonstrate that there are two kinds of electric charges, which Franklin called *positive* and *negative*. Positive charges appear when one rubs a glass rod with silk, whereas negative charges arise after rubbing a rubber rod with fur. If one brings a positively charged glass rod near a negatively charged rubber rod, suspended by a nonmetallic thread, then the rubber rod will move toward the glass rod. Conversely, two rods having the electric charges of the same sign will repulse each other. Therefore, it is possible to make a very important conclusion known today as a *basic law of electrostatics*, stating that like electric charges repel one another and unlike charges attract one another.

Knowing the law of electrostatics, Franklin formulated the *principle of conservation of electric charge.* According to this principle, the rubbing of one body against the other does not produce electric charges but rather *transfers* them from one body to another. Additionally, Franklin demonstrated that a transfer of electric charges from one body to another can also be accomplished by induction. Consider a neutrally charged metallic ball (Fig. 1.3.1). This ball charges electrically when a charged rod is placed near it, without any physical contact. In this case half of the ball will have a positive charge and the other half negative.

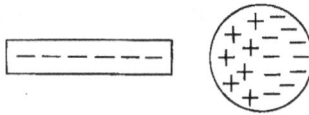

Figure 1.3.1. Charging of a metallic ball by induction.

Franklin explained that electric charge appears because of either excess or deficiency of the so-called "electric fluid." His idea of single electric fluid successfully competed against the idea of two electric fluids advanced by several prominent scientists of that time. Later he astonished the world by discovering a device that was capable of accumulating an electric charge. Franklin spent much of his time studying the nature of lightning. By conducting several ingenious experiments, Franklin proved convincingly that the lightning discharge is an electrical phenomenon.

Charles Coulomb (1736-1806) –To discover a quantitative basis for Franklin's laws of electrostatics, the French physicist Charles Coulomb conducted a series of experiments. Using the torsion balance, a device of his own his invention, he measured the twist in a suspended fiber. He then used this technique in his experiments with magnets. Based on the results of these experiments, Coulomb established the Law of Electric Force F_e between two charged particles that states the following:

The electric force between two charged particles F_e is directly proportional to the product of the charges e_1 and e_2 of the particles and is inversely proportional to the square of the distance r between these particles, i.e.:

$$F_e = \frac{e_1 e_2}{4\pi\varepsilon_0 r^2}$$

where ε_0 is the electric constant.

Coulomb obtained a similar equation in respect to magnetic charges, but he claimed that there were no direct relationships between electricity and magnetism. It was obvious to him that his law of electric force F_e was very similar to Newton's law of gravitational force F_g acting between two bodies with gravitational masses m_1 and m_2 separated by the distance r. To emphasize the similarity between these two laws, Coulomb used the term "electrical mass." But, unlike Newton's law that dealt with only attraction forces, his law applied to both attraction and repulsion forces.

André-Marie Ampère (1775-1836) –The discovery of electromagnetism by the Dutch physicist Hans Oersted in 1820 became a turning point in the scientific career of the French physicist and mathematician André-Marie Ampère. Jointly with the French physicist Augustin Fresnel (1788-1827), Ampère proposed the theory of electromagnetic molecule to explain why a current-carrying helical coil behaved like a magnet.

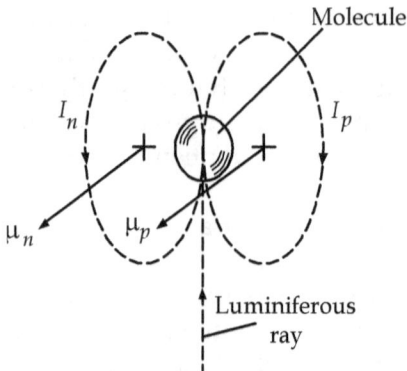

Figure 1.3.2.
Ampère's
electrodynamic
molecule.

Their idea was that myriads of small circular electric currents existed within the molecules of iron. Once these molecules were aligned, the resultant current would act precisely as the concentric currents in the helical coil. Both positive and negative fluids pour out of the top, flow around the molecule and reenter at the bottom (Fig. 1.3.2). Since the positive and negative fluids I_p and I_n are circling around the molecule in opposite directions, their resultant magnetic moments μ_p and μ_n are pointed in the same direction, thus doubling the magnetic moment of the molecule. Ampère and

Fresnel proposed that the molecular electric currents in iron were induced by the luminiferous ether that pervades space and penetrates matter.

After describing the noumena of the electromagnetism, the next step for Ampère was to apply mathematics to the phenomena. He successfully completed this task in 1820 by deducing his famous relationship that became known as *Ampère's law*. This law establishes a relationship between the magnetic induction B along a closed path around a wire and the current I in the wire. If the magnetic path around the wire with radius R is circular, then the magnetic induction B at any distance r from the center of the wire is equal to (Fig. 1.3.3):

$$B = \frac{\mu_0 I}{2\pi r}$$

where μ_0 is the magnetic permeability of free space.

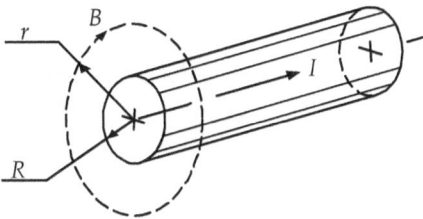

Figure 1.3.3.
Magnetic induction B
around a wire
with current I.

Michael Faraday (1792-1867) – The British chemist Michael Faraday made a great contribution to science by his discoveries in electromagnetism. His inventions laid out the foundation for the construction of electric motors, generators and transformers based on this phenomenon. Faraday began his discoveries in electromagnetism by making the same very logical proposition that several other scientists advanced at the time.

If magnetism is produced by electricity, as was clearly demonstrated by Ampère, then the reverse phenomenon should also exist and electricity must be produced by magnetism. To prove this proposition, Faraday conducted several experiments. In one of his most interesting experiments, Faraday used what became known as Faraday's induction ring, a simple device that resembles a modern transformer (Fig. 1.3.4). The ring was made of soft iron and contained two coils wound around it. Faraday connected one coil first to a galvanometer, and the other coil to the battery through a switch. After connecting the battery, he noticed that the needle of the galvanometer moved and then came back to zero. The needle moved again but in the opposite direction when Faraday disconnected the battery.

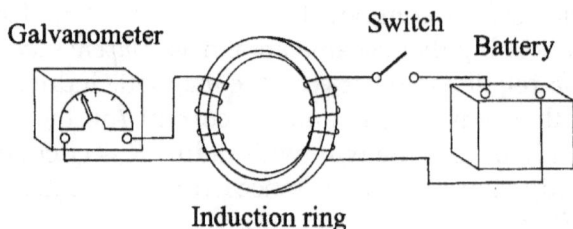

Figure 1.3.4. Faraday's experiments with electromagnetism.

It was clear now that the electric current appeared only when the magnetic field was changing. In his presentation to the Royal Institution, Faraday summarized his historic discovery in a single statement:

> Whenever a magnetic force increases, it produces electricity; the faster it increases or decreases, the more electricity it produces.

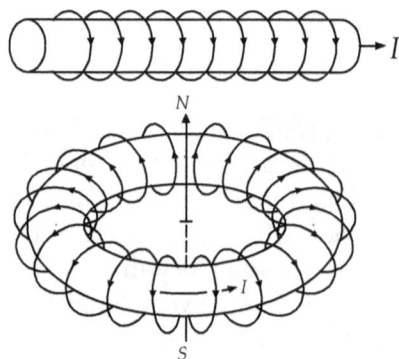

Figure 1.3.5. Faraday's lines of force in a straight wire and in the loop, both carrying electric current.

To illustrate the electromagnetic effects, Faraday employed *lines of force*. A straight current-carrying wire was surrounded by circular lines of force (Fig. 1.3.5). After bending this wire into a loop, the lines of force would become distorted, so a greater concentration of the lines of force would be created inside the loop. This concentration of the lines of force, according to Faraday, produces magnetic forces with the north and south poles located at opposite sides of the loop.

Faraday's philosophical way of thinking inevitably led him to the conviction that all known physical phenomena could be explained by one unified theory. He wondered if besides the electric nature of magnetism the nature of gravity is also electrical. He was very familiar with the idea that gravity may be explicable as the result of slight deviations of celestial bodies from the state of electrical neutrality. Father Giovan Battista (1716-1781), known in the secular world as Francesco Beccaria, originally proposed this fascinating idea in the mid-eighteenth century. Since that time, several scientists, including the German natural philosopher Franz Aepinus (1724-1802) and the Italian physicist Ottaviano-Fabrizio Mossotti (1791-1863), brought to light the idea that the electrostatic forces of attraction between unlike charges slightly outbalance the repulsion forces between like charges.

As was usual in application to any other phenomenon, Faraday began his investigations of the relationship between gravity and electricity by conducting experiments. This time, unfortunately, all of his experiments failed. Insufficient resolution of his instruments, Faraday thought, did not permit him to detect the infinitely small deviation from the state of electrical neutrality that was needed to produce gravity. Nevertheless, he was strongly convinced that, in his words,

Universal gravitation is mere residual phenomenon
of electrical attraction and repulsion.

Wilhelm Weber (1804-1891) – Nowadays, we describe *relativity* as the dependence of the physical properties of objects on their velocity. The German physicist Wilhelm Weber was among the first scientists who described this phenomenon. His long scientific life was both interesting and contentious. In 1846 he formulated his controversial *law of electric force*. According to this law, the electric force F_e between two moving electric charges e_1 and e_2 separated by the distance r is equal to:

$$F_e = \frac{e_1 e_2}{4\pi\varepsilon_0 r^2}\left(1 - \frac{V^2 + 2ra}{2c^2}\right)$$

where
   V = radial velocity of electric charges in respect to each other
   a = radial acceleration of electric charges in respect to each
      other
   c = velocity of light.

The dominant term in the above equation is Coulomb's force $e_1e_2/4\pi\varepsilon_0 r^2$. The remaining terms (in brackets) modify this force, making it either attractive or repulsive, when the charges are in motion relative to each other. Weber was the first scientist to propose that the *interaction between particles is velocity-dependent.* Another interesting consequence of Weber's law was that two particles of the same sign of charge can revolve around each other on stable orbits. If this is true, Weber speculated, then his law might lead to an understanding of the laws of light and heat radiation. Not everyone accepted Weber's theory. His strongest opponent was the German physicist Hermann von Helmholtz (1821-1894), who believed that Weber's law of electric force violated the principle of conservation of energy.

In 1882 Weber developed further the ideas of the German astrophysicist Johann Zöllner (1834-1882). According to Zöllner, all matter is compounded of electrically charged particles, held together in various stable configurations by the action of the Weberian law of force. Even gravitation was interpreted by adopting in essence the earlier hypotheses of Aepinus and Mossotti that the attraction electrostatic forces between unlike charges slightly outbalance the repulsion forces between like charges.

William Thomson (1824-1907) - The British physicist William Thomson (Lord Kelvin) is mostly known for his works in thermodynamics. He also turned out to be one of the most ardent supporters of the vortex theory. Thomson argued that from any galvanic current there extends a moving spiral that coils about the line of magnetic forces, which passes through the center of the axis of the current (Fig. 1.3.6). He proposed that the current itself consists of the trapping of a segment of this spiral in ponderable matter (matter that has weight). When extending this idea to light waves, he believed that transverse vibrations of the particles of the moving spirals caused the light waves to propagate. He pictured that the plane of polarization of the waves is rotated depending upon the direction of the magnetic force.

Thomson hoped to explain all electromagnetic phenomena with this representation. For instance, the twisting motion of the spiraling helices could produce magnetic effects. These helices, when located in matter, would become currents. In his vortex theory Thomson treated the luminiferous ether as a medium for the propagation of waves; light waves were viewed as oscillations of ether. Present everywhere, even in empty space, ether was massless and at the same time an elastic solid medium.

The last requirement was imperative to account for the major properties of light. Furthermore, in order for reflection and refraction of light to occur, the ether had to be absolutely incompressible.

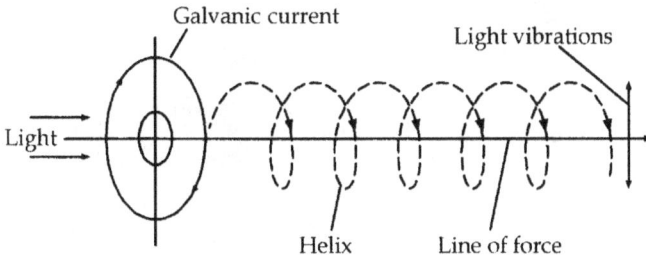

Figure 1.3.6. William Thomsoson's presentation of the electromagnetic wave. Adapted from C.C. Gillispie (1971).

Thomson made his most interesting discoveries in the vortex theory in collaboration with the Scottish physicist Peter Tait (1831-1901). Thomson renewed his interest in spirals after reading a 1858 paper by Helmholtz. In his work, Helmholtz demonstrated that the motions of linear fluid vortices exhibit striking patterns: two adjacent figures appear to repel one another in a complex manner. This effect is the result solely of pressures produced in the medium; once established by some unknown means in a non-viscid medium, the vortices cannot be destroyed by any mechanical process and, thus, are made eternal.

Tait suggested to Thomson that the same concept could be applied to explain the structure of light and atoms, and proposed the idea of a demonstration. He also designed the apparatus for the demonstration. It consisted of a couple of boxes, two pieces of cloth, and a few bottles of chemicals. A sturdy cloth covered one side of each box, while the opposite side had a circular hole. Within each box, Tait mixed the vapors of an acid with ammonia to produce thick clouds of smoke. When he struck firm blows on the cloth, numerous circular rings shot out of the hole. The members of the Royal Society of Edinburgh were amused by the behavior of the circular rings, which filled the entire room with smoke rings bouncing off each other. The rings shook violently from the collisions, making the impression that they were made of rubber. It was impossible to cut these rings with a knife; they simply moved away from the blade.

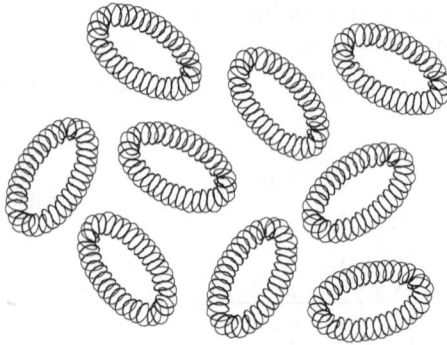

Figure 1.3.7. Tait's smoke rings as toroidal vortices.

When looking closely at the smoke rings (Fig. 1.3.7), one could clearly recognize the neatly wound toroidal spirals. Thomson and Tait used this demonstration to point out that these vortex rings of smoke behaved just like atoms, and that all the properties of atoms stemmed from vortex spin.

James Clerk Maxwell (1831-79) - The British physicist James Clerk Maxwell (1831-79) is best known for the development of the theory of electromagnetism. Very few of us, however, know that during the many years of his scientific life he was a great supporter of the vortex theory. He worked closely with Faraday, Thomson and Tait. In 1862, Maxwell devised a model to illustrate Faraday's law of electromagnetic induction. He proposed an analogy between lines of force in the electromagnetic field on the one hand, and vortex filaments in a liquid with finite fixed boundaries on the other. The following year he employed vortex theory to explain the rotation of the plane of polarization of light, traveling along the force in magnetic field. He envisioned a beam of linearly polarized light as consisting of two circularly polarized components rotating in opposite directions. The vortices, created by the magnetic field, speeded up the rotation of one component and slowed down that of the other. The outcome was a rotation of the plane of polarization.

In 1864 Maxwell suddenly abandoned the vortex theory. This became evident after he published his famous paper *The Dynamic Theory of the Electromagnetic Field* in that year. In our time this paper is considered to be the cornerstone of the theory of electromagnetism. The essential point of this theory was that the electromagnetic field is transverse vibrations, supported by electromagnetic medium and propagated at the velocity of light. In his theory, Maxwell did not use either the vortices or, in fact, any other physical model, allowing the mathematical equations alone to do the job. The publication of this paper brought Maxwell neither fame nor glory. Moreover, because of this publication, he alienated two of his closest friends in the entire scientific community, Faraday and Thomson.

In fairness to Maxwell we shall remember that before publishing this paper, he did his best to explain electromagnetism by using the vortex theory, to which he himself became attached wholeheartedly. He tried to model the electric current in the electromagnetic wave by employing some sensible physical analogies. This current was badly needed because, according to Faraday's law of electromagnetic induction, the magnetic field in the electromagnetic wave can only be created if there is a change in the electric current. However, to introduce the electric current, which he called the *displacement current*, the electromagnetic wave must contain a device similar to a capacitor that is periodically charged and discharged. Certainly, there was no way of finding such a capacitor in the colorless and weightless electromagnetic wave.

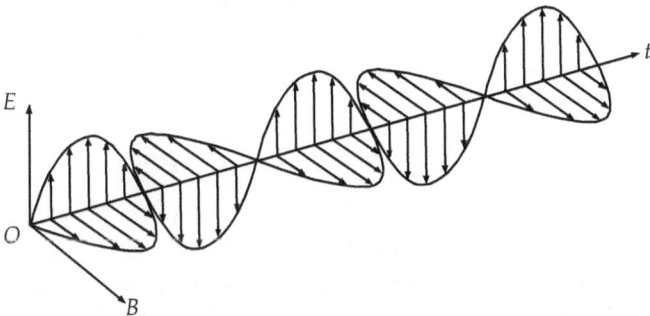

Figure 1.3.8. Maxwell's presentation of the electromagnetic wave.

Frustrated, Maxwell eventually dropped all his attempts to model the electromagnetic wave by employing any physical analogy, and presented his theory with a set of mathematical equations. The good news was that this purely mathematical theory worked. It explained how a sinusoidal electric field E and a sinusoidal magnetic field B, oscillating in two mutually perpendicular planes, were generating each other and forming an electromagnetic wave (Fig. 1.3.8). It also stipulated that the velocity of the propagation of the electromagnetic wave was equal to the velocity of light.

Here again, Maxwell had a serious problem. Unable to derive the above equation from his vortex model, he derived it by making some *ad hoc* assumptions regarding the elastic structure of the medium. Despite strong opposition to his theory, Maxwell published his *Treatise* in 1873. Its publication, however, did not bring him any personal satisfaction: he failed again to explain his equations from a physical point of view and had to drop any sensible reference to any kind of a physical model.

He recognized his failure and was afraid that some young scientists would see in his *Treatise* a sign that there was no longer a need for discovering a physical meaning of the mathematical equations that describe physical phenomena as long as these equations predicted accurately the results of the experiments. Maxwell was concerned that this viewpoint would eventually lead to the complete dissociation of physics from physical models that educated laymen could understand. Therefore, when Maxwell started on a new edition of his *Treatise* in 1879, his goal was to restore Faraday's traditions and to rewrite his theory by employing a clear physical model. Unfortunately, he was not able to meet his goal. The great scientist died of cancer on 5 November 1879.

1.4 Atomic & Nuclear Age

The atomic and nuclear age is associated with experimental discoveries of particles of matter and radiation on atomic and subatomic levels. Two new theories, quantum mechanics and the theory of relativity, provided a strong theoretical base for understanding the structures of elementary particles and the forces acting between them.

Max Planck (1858-1947) – The German physicist Max Planck originated quantum theory, making the year 1900 a transitional point between classical and modern physics. Planck's path to quanta, his greatest discovery, began with his studies of one of the problems of thermodynamics known as the black body radiation. The ideal black body absorbs all radiation that incidents on it.

The best approximation of a black body is a container with a small hole in it, into which radiation shines and is trapped inside (Fig. 1.4.1). As light enters the container through the hole, a part of it is reflected and another part is absorbed on each reflection from the interior walls. After many reflections, essentially all of the incident energy is absorbed.

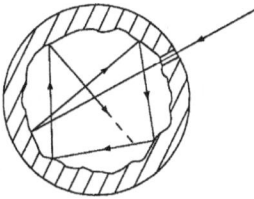

Figure 1.4.1.
An approximation
of a black body
radiation.

To derive the equation that worked for all wavelengths, Planck had to assume that radiation is emitted or received in energy packets called *quanta* with the quanta energy E equal to a product of the *Planck constant h* and the wave frequency f:

$$E = hf$$

This was the beginning of the quantum mechanics that was further developed by several physicists in the 20th century.

Albert Einstein (1879-1955) - In 1905 the German-Swiss-American theoretical physicist Albert Einstein successfully used Planck's new theory in his prediction of photoelectric effect. Simply, photoelectric effect is the release of electrons from a substance under the influence of light or other electromagnetic radiation. In his 1905 paper, Einstein demonstrated that the particles of light would have to be such that the ratio of energy E to frequency f must be equal to the Planck constant h. Thus, according to Einstein, the concept of quanta is applicable not only to black body thermal radiation, but also to light.

During that same year Einstein introduced his Special Theory of Relativity. Before Einstein published this theory, people calculated very simply the relationships between two objects moving with constant velocities. Suppose, for instance, that you are traveling in a train that moves with velocity V_1 in respect to a train station, and you are walking towards the front of your carriage with velocity V_2 in respect to the carriage floor. What is your velocity in respect to the train station? A logical answer is $V_1 + V_2$. And this is also a correct answer according to the Galilean law of addition for velocities.

Einstein showed that the law of addition for velocities becomes completely different when one of the objects is a photon moving with the velocity of light. Coming from his principle of relativity was one of the most intriguing propositions of his time:

In empty space light travels with the definite velocity c
that does not depend on the motion of its source.

The above proposition obviously clashes with the Galilean law of addition for velocities. For instance, if a source of light travels in respect to a stationary object with the velocity V_1, then, according to Galilei, the light coming from this source will travel in respect to the stationary object with velocity $V_1 + c$, and not c as followed from the above proposition. Einstein resolved this contradiction by employing the Lorentz-Fitzgerald Transformation derived earlier by the Dutch theoretical physicist Hendrik Lorentz (1853-1928) and the Irish physicist George FitzGerald (1851-1901) to prove the existence of ether. According to these equations, the following changes occur with a moving object:

- The observed length of an object decreases with increase of velocity in comparison with its length at rest.
- The observed mass of an object increases with increase of velocity in comparison with its rest mass.
- According to a stationary observer, a moving clock runs slower than an identical stationary clock. This effect is known as *time dilatation*.

Einstein, however, using this known relativistic equation, provided a new physical meaning for it. The outburst of Einstein's new ideas in 1905 did not end with the publication of the special theory of relativity. In his fifth paper, published in November of that year, Einstein showed by calculation that if a body gives an amount E of energy in the form of light, then its mass decreases by an amount E/c^2. Thus, mass and energy are wholly equivalent. In 1907 Einstein presented this idea in a long and mainly expository paper that contained his famous equation, establishing the relation between energy E and mass m:

$$E = mc^2$$

In 1915 Einstein introduced his General Theory of Relativity in which he applied the concept of *curved spacetime* to explain gravity. The idea of curved space was introduced earlier by several great scientists, including the Russian mathematician Nikolai Lobachevsky (1793-1856), the German mathematician Karl Friedrich Gauss (1777-1855), the Hungarian mathematician János Bolyai (1755-1856) and the German mathematician Georg Riemann (1826-1866).

The term "spacetime" was originally proposed in 1907 by the Russian-German mathematician Hermann Minkowski (1864-1909). He used this term to describe a coordinate system of the four-dimensional space. It was an important step in breaking away from the Newtonian concept of absolute time and space. When time and space are considered independent of each other, the location of an object in a three-dimensional space is defined by three dimensions x, y, and z.

To connect space with time t, Minkowski added one more dimension ict in which $i = \sqrt{-1}$ is the imaginary number. Thus, the modulus of the vector u defining the location of a point on a curved space was defined by the equation:

$$u^2 = x^2 + y^2 + z^2 - (ct)^2$$

Einstein expanded the application of the term spacetime in his General Theory of Relativity by using a special branch of mathematics called *tensor calculus*. The result was a groundbreaking departure from the Newtonian interpretation of gravity. Gravitation was no longer a force but an intrinsic curvature of spacetime. It was the curved spacetime around the Sun that attracted the planets. In curved spacetime the motions of the planets were represented by the shortest distances, called *geodesics*.

Numerous experiments conducted since 1915 proved that Einstein's theory of general relativity was correct. This theory had also found its application in the contemporary astrophysics. According to his theory, the spacetime can in principle be warped so strongly by a huge mass that any radiation emitted from the mass curves back in again and cannot escape. These huge masses are thought to exist as black holes.

Theodor Kaluza (1873-1916) – In 1918 the German mathematician Hermann Weyl (1885-1955) proposed an extension of Einstein's general theory of relativity. He modified Einstein's spacetime by providing one additional geometric parameter for electromagnetism, making this parameter an equal partner to gravitational curvature. Einstein rejected Weyl's idea, but, in 1919, the German mathematician and physicist Theodor Kaluza took a completely different approach. He managed to leave Einstein's equations unchanged with only one principal modification: these equations were now applied to five dimensions instead of four. This fifth dimension was expected to account for electromagnetism.

Kaluza presented his multidimensional space by wrapping a space of a higher dimension around a space of a lower adjacent dimension. Consider that a few hundred feet of garden hose is stretched across a canyon, and you view it from a far distance away. From this distance the hose will appear as a one-dimensional string. Therefore, an ant located on the hose would have only one dimension in which to walk: the left-right dimension along the hose's length.

From a magnified perspective, we will see that besides the left-right dimension along the hose's length there is the second dimension along which the ant can walk, either clockwise or counterclockwise. We may say that the second dimension of the hose as "wrapped up" around the one-dimensional material. Similarly, it is possible to describe mathematically a three-dimensional hose "wrapped around" the two-dimensional material, and so on and so forth. So, how in the world did Kaluza manage to make Einstein's four-dimensional equations work in a five-dimensional spacetime? The secret was in his technique called *compactification* that allows one to reduce a spacetime of a higher dimension to a spacetime of a lower dimension.

In his wonderful book *Q is for Quantum*, John Gribbin provides us with an excellent example of how compactification works. One can make a three-dimensional space in the form of a spiral pipe from a thin sheet that may be viewed as a two-dimensional space. From a distant vantage point the three-dimensional spiral pipe will look like a one-dimensional line. Similarly, it is possible to describe mathematically a five-dimensional spiral pipe made of a four-dimensional material. There is one condition, though: the "wrapping" must happen on a scale much smaller than that of an atomic nucleus. In 1926, independently of Kaluza, the Swedish theoretical physicist Oskar Klein (1894-1977) refined Kaluza's theory to take into account the requirements of quantum theory. Although the two scientists never worked together, their discoveries became known as the *Kaluza-Klein theory.*

Karl Schwarzchild (1873-1916) – The idea of the existence of black holes was first proposed in 1783 by the British astronomer John Mitchell (c. 1724-1793). If the stars were sufficiently massive and compact, he thought, then light would not be able to escape from the surface of the "dark stars," as the black hole was called at that time. He further suggested that the dark stars might be detectable through their gravitational effect on nearby objects. But his idea was quickly forgotten. It was revived 133 years later by the German astronomer Karl Schwarzchild. In the last year of his life he wrote two papers in which he gave the first solution to the complex partial differential equations of Einstein's general theory of relativity.

His calculations showed that for any object squeezed inside a sphere with a critical radius, now called the *Schwarzchild radius,* space would curve around the object and separate it from the rest of the universe. It would become a self-contained universe, from which nothing (*not even light*) could escape. The Schwarzchild radius R corresponding to any mass m is equal to:

$$R = \frac{2mG}{c^2}$$

According to the above equation, the Earth would become a black hole if compressed to a sphere with the radius $R = 0.88$ cm, while the same would happen to the Sun, when $R = 2.9$ km. Still, in spite of Einstein's support for Schwarzchild's theory, the idea of a black hole remained a mere theoretical curiosity for half a century. It had to wait for conventional ground-based telescopes to be supplemented with the Hubble Space Telescope before astronomers could obtain indisputable observations to confirm the existence of black holes.

Edwin D. Babbitt (1828-1905) - Edwin D. Babbitt was one of the most influential American pioneers and writers in the field of color research and color therapy. In 1878 he published his *The Principles of Light and Color,* in which he outlined spiral structure of an atom that he called *anu.* Babbitt used his structure of this atom to explain bonding, electricity, light, color, friction, psychic power, and nearly everything else.

Charles Leadbeater (1847-1934) – In 1909, a renegade Anglican clergyman Charles Leadbeater and his assistance Mrs. Besant published the book titled *Occult Chemistry*. In their book they described the basic subatomic structure as seven layers of recursive spirals around other spirals.

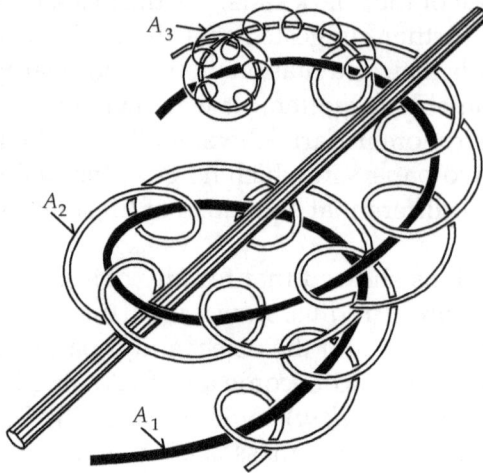

Figure 1.4.2. Leadbeater's multiple-level spiral.

As shown in Figure 1.4.2, a winding of a spiral A_1 is wound by a spiral A_2 that, in its turn, is wound by a spiral A_3. Similarly, spiral A_3 would be wound by spiral A_4, and so on up to spiral A_7. Although the atomic structures described by Leadbeater and Besant did not make much sense to their contemporaries, these structures became recognizable in some atomic theories developed several decades later. The British physicist Stephen Phillips published, in 1980, a book titled *Extra-Sensory Perception of Quarks* in which he presented the work of Leadbeater and Besant in light of modern theories of atomic structure.

Walter Russell (1871-1963) - Another outstanding amateur self-educated scientist of the 20th century was the American artist and sculptor Walter Russell summarized the results of his revelation in his two masterfully written books: *The Universal One* that was originally published in 1927 and *The Secret of Light* published in 1947. Based on the concept of vortices, he attempted to explain the nature of light, electricity, and magnetism. He went as far as to devise his own Table of the Elements and to predict several new elements unknown during his time. After reading his books, one may only wonder how a man with no formal education in any branch of sciences was able to penetrate intuitively the depths of understanding the secrets of nature. The following quotation from Walter Russell was provided by Glenn Clark in his enlightening book *The Man Who Tapped the Secrets of the Universe:*

> That the man calls matter, or substance, has no existence whatsoever. So-called matter is but waves of the motion of light, electrically divided into opposite pairs, then electrically conditioned and patterned into what we call various substances of matter. Briefly put it, matter is the motion of light, and motion is not substance. It only appears to be. Take motion away and there would not be even appearance of substance.

Ernest Rutherford (1871-1937) – The New-Zealand-British physicist Ernest Rutherford is considered by the scientific community to be the founder of nuclear physics. He is mostly known for his discoveries in radioactivity and atomic structure. His research in radioactivity was triggered by the discovery of X-rays by the German physicist Wilhelm Röntgen (1845-1923) in 1895, the discovery of the radioactivity of uranium by the French physicist Henri Becquerel (1852-1908) in 1896, and the discovery of two new radioactive elements, polonium and radium, made by two French physicists, Pierre Curie (1859-1906) and Marie Skłodowska-Curie (1867-1934) in 1898.

Based on his extensive experimental research on radioactivity, jointly with Frederick Soddy, Rutherford developed the theory of radioactivity. The theory explained how the decay of radioactive atoms produces the transmutation of a parent into a daughter element. It also identified the role of the three processes: alpha (a) decay, beta (β) decay, and gamma (γ) decay. During the alpha decay, a nucleus emits a heavy alpha particle that turned out to be a nucleus of helium. The beta decay involves emission of light charged particles that later were recognized as the electrons and positrons. The gamma decay involves the radiation of high-energy gamma rays. This work brought Rutherford the Nobel Prize in chemistry that he received in 1908.

Rutherford's most sensational discovery came out of the so-called *scattering experiments* that he conducted in 1909. Actually, as a boss and a Nobel Prize winner, he did not have to conduct these tedious experiments by himself, and assigned them to two of his most trusted underlings: a postdoc, Hans Geiger, and an undergraduate student, Ernest Marsden. The main purpose of these experiments was to investigate atomic structures of a metallic samples made of thin foils. A beam of fast moving alpha particles, emitted by a radioactive element, was first passed through a diaphragm and then shot at a thin foil. In passing through the foil, the alpha particles collided with the constituent atoms and some of the particles were scattered in different directions. In other words, the alpha particles, after hitting something inside the metal foil, bounced back. Rutherford rightfully saw the only reason for such a dramatic behavior of the positively charged alpha particles: they were reflected by an even heavier positively charged particle. The nucleus of an atom was now discovered.

In the Rutherford's model of the atom, the positive electric charge was concentrated inside a small nucleus located at the center of the atom, and the negatively charged electrons were moving around it, resembling the solar system. In spite of its attractiveness, the new model soon ran into a serious problem. According to the classical theory of electromagnetic emission, the negatively charged electrons circling the nucleus must emit light waves, carrying with them a part of the electron's energy. Consequently, the electrons would completely lose their energy and spiral into the nucleus. It was a disaster, but nobody could find an immediate solution. The impasse had lasted for several more years before it was resolved, thanks to the contributions made by the three outstanding scientists, the British experimental physicist Henry Moseley, the Danish theoretical physicist Neils Bohr, and the German physicist Arnold Sommerfeld.

Henry Moseley (1887-1915) - During his short life, the British experimental physicist Henry Moseley made the first step toward a better understanding of Rutherford's model by discovering the physical meaning of electric charges of atomic nuclei for different chemical elements. He also provided a completely new interpretation of the periodic table of chemical elements originally developed in 1869 by the Russian chemist Dmitri Mendeleyev (1834-1907). In this table, the elements' relative masses were the principal factors in determining the periodicity of the properties of the chemical elements. Nobody questioned the validity of Mendeleyev's Periodic Table of the Elements for 44 years. This all changed in 1913 after Moseley completed a series of his extraordinary experiments with the X-ray spectra.

In his thorough experiments, Moseley used over thirty different metals (from aluminum to gold) as targets. He found that the
X-ray spectra lines changed regularly in position from element to
element exactly in accordance with Mendeleyev's Periodic Table.
Moseley said of his results: "We have here a proof that there is in
the atom a fundamental quantity, which increases by regular steps
as we pass from one element to the next. The quantity can only be
the charge on the central positive nucleus." After Moseley's discovery, the electric charge of the nucleus replaced the element's
relative mass as the principal factor determining the periodicity of
the properties of the chemical elements.

Niels Bohr (1885-1962) – In 1913 the Dutch physicist Niels
Bohr proposed a model of the hydrogen atom. This was a significant improvement of the model of the hydrogen atom proposed
earlier by Ernest Rutherford. Bohr adopted Rutherford's basic
premise that in the hydrogen atom an electron orbits around a
proton. But, Bohr rightfully wondered, why would the electron
remain stable in its path around the proton? Fortunately, he had
at his disposal three pieces of knowledge that were critical for
finding the answer to this question. Bohr was certainly familiar
with the first two: Planck's discovery of quanta and Einstein's
theory of photoelectric effect. The third crucial piece of knowledge came to Bohr from the results of investigation of atomic
spectra of the hydrogen atom conducted by Balmer and Rydberg.
The phenomenon of atomic spectra is rather simple. When an
electric charge passes through a gas and the resulting light is dis-
persed by a prism, a spectrum of sharp lines of discrete wavelengths, or atomic spectra, is observed. The line spectrum for
atomic hydrogen was the simplest one, because its lines had a dis-
tinct regularity of spacing.
In 1884 a Swiss school teacher, Johann Jacob Balmer (1825-
1898), found a formula that correctly predicted the wavelengths
λ_{mn} of four visible lines of hydrogen H_α, H_β, H_γ, and H_δ that became known as *Balmer series* (Fig. 1.4.3). The Swiss spectroscopist
Johannes Robert Rydberg (1854-1919) further extended Balmer's
work. Working on atomic emission spectra in 1890, Rydberg
found a simple general equation for the frequencies f_{mn} of some of
the spectra lines for the hydrogen atom:

$$f_{mn} = R_\infty c \left(\frac{1}{m^2} - \frac{1}{n^2} \right)$$

where
R_∞ = Rydberg constant
m, n = integer numbers.

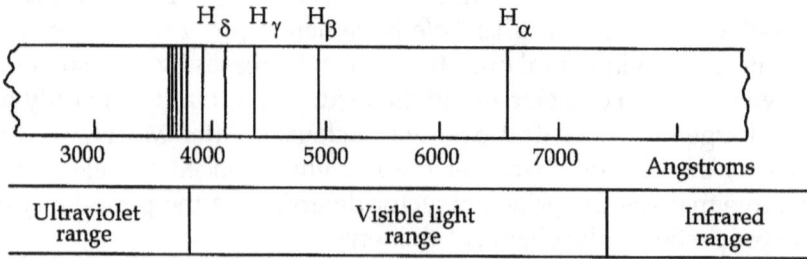

Figure 1.4.3. The line spectrum for atomic hydrogen.
Adapted from G. Gamow (1985).

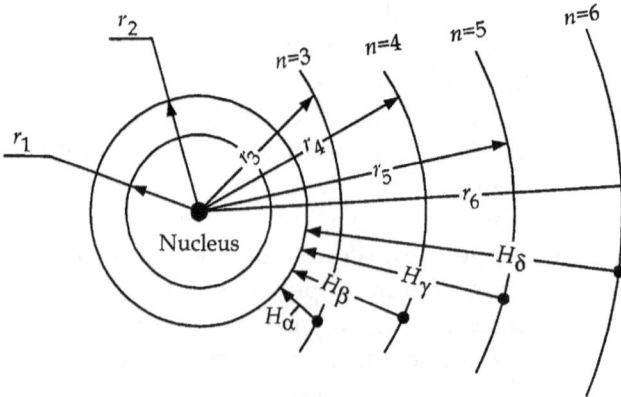

Figure 1.4.4. Electron orbits and spectra lines of a hydrogen atom.
Adapted from H.E. White (1934).

In Bohr's model of the hydrogen atom (Fig. 1.4.4), the electron was stable only on those orbits in which its orbital angular momentum p_a was related to Planck constant h by the equation:

$$p_a = mVr = \frac{nh}{2\pi}$$

where m = electron mass, V = electron orbital velocity, r = radius of electron orbit, and n = quantum energy states 1, 2, 3, . . .

The above formula allowed one to calculate the energies (and frequencies) of the photons that were emitted, when the electron was transferred from higher to lower energy states.

Arnold Sommerfeld (1868-1951) – The German physicist Arnold Sommerfeld further elaborated Bohr's atomic model. The purpose of his innovation was to account for the so-called *fine structure* of the spectra lines that appeared in close proximity to the main spectra lines H_α, H_β, H_γ, and H_δ.

His solution was twofold. Firstly, he proposed that for all orbital quantum states that are greater then the ground state $n = 1$, there must be elliptical electron orbits in addition to the circular orbits described by Bohr's model. As shown in Fig. 1.4.5, in the hydrogen atom there is only one circular electron orbit corresponding to the ground state with the principal quantum number $n = 1$. For $n = 2$, there are four orbits, one circular and three elliptical. For $n = 3$, there are nine orbits, one circular and eight elliptical. To assure stability of electrons on all these orbits, it was necessary to add two additional quantum numbers: l (orbital quantum number) and m_l (orbital magnetic quantum number). Without this additional quantization all electrons would end up at the ground energy state.

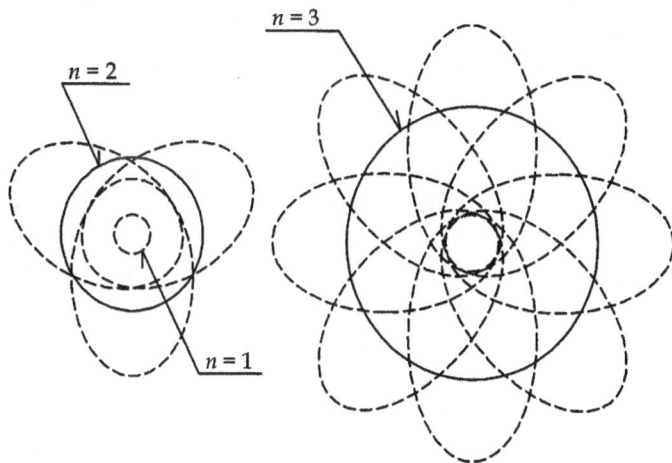

Figure 1.4.5. The quantum electron orbits in the hydrogen atom.

Sommerfeld showed that the ellipticity should result in a relativistic effect and discovered that the frequency difference between the fine structure components depends on the so-called *fine structure constant* α given by the equation:

$$\alpha = \frac{e^2}{2\varepsilon_0 hc}$$

Remarkably, the fine structure constant a is dimensionless and approximately equal to the inverse of 137. The fine structure constant showed up later in many models of elementary particles proposed by other physicists. Because the constant does not have dimensions, some scientists believe that it has a much greater significance in nature than the constants with dimensions invented by people.

1.5 Models of Elementary Matter Particles

In spite of the great progress made in particle physics during the 20th century, a simple question still remained unanswered: "What is the structure of an electron?" Can it be described by considering it as either a wave or a particle? The presentation as a wave required an explanation of how the wave properties of an electron can be related to its physical properties. If the electron is a particle of a certain size, then how can it withstand the repulsion negative electric forces that would try to rip it apart? The quantum theory "solved" the latter problem superficially by presenting the electron as a mathematical point having no dimensions.

Some physicists did not like the presentation of an electron as a charged point mass. Among these scientists was the German physicist Max Born (1882-1970). He was troubled by the fact that electron energy per unit volume becomes infinite when its radius is equal to zero, a paradox he could not accept. According to his calculations based on experimental data and Maxwell's theory of electromagnetism, the radius r_e of an electron with rest mass m_e and rest charge e is equal to:

$$r_e = \frac{0.408 e^2}{\pi \varepsilon_0 m_e c^2}$$

In 1903, the German physicist Max Abraham (1875-1922) developed a model of the electron, in which he assumed the electron to be a *rigid sphere*, with a uniform surface charge distribution. He also assumed that the electron's mass originated entirely in its own electromagnetic field. Based on these assumptions, he was able to determine that the radius r_e of the electron is equal to:

$$r_e = \frac{0.125e^2}{\pi \varepsilon_0 m_e c^2}$$

A different radius of the electron carries the name of the Dutch theoretical physicist Hendrik Lorentz (1853-1928) and is known as the Lorentz radius of an electron. It is equal to:

$$r_e = \frac{0.25e^2}{\pi \varepsilon_0 m_e c^2}$$

Neither of the above models, however, accounted for the electron's spin (angular momentum) and magnetic moment, the two extremely important properties of an electron that at that time had not yet been discovered. In any case, these models allowed the electron to have only one degree of freedom, the sphere radius, which limited the models' capability to describe the various properties of the electron.

Alfred Parson (1889-1970) - In 1915, the British chemist and physicist Alfred Parson proposed a more sophisticated model of an electron, in which the electron charge was distributed over the surface of a *toroidal ring* (Fig. 1.5.1). The ring was very thin, with the radius r of about 5×10^{-11} m, much smaller than the overall ring radius R. The ring produced an overall magnetic field ("spin") due to the current created by the charged elements moving along the ring with the velocity of light c.

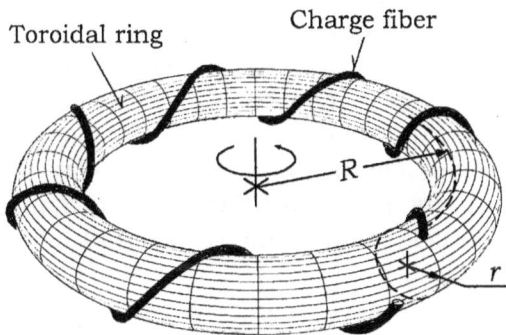

Figure 1.5.1. Parson's toroidal ring model of an electron.

In this configuration, the charged elements circled the center of the ring, but the ring as a whole did not radiate, because both electric and magnetic fields remained constant. The model treated electrons or protons as bundles of "fibers" or "plasmoids" with the total charge equal to the elementary charge $\pm e$. The fibers twisted around the toroidal ring as they wrapped around its surface, forming a slinky-like helix. Each helical fiber was twisted around the ring an integer number of times to account for quantum values of angular momentum and radiation. The number of fibers was odd. The helicity of the twist distinguished the electron from the proton. Reportedly, the model explained how particles linked together to form atoms.

In 1956, the American physicist Winston Bostick (1916-1991) demonstrated experimentally the existence of so-called *plasmoids*, force-free charged strings. Ten years later he proposed a model of an electron composed of helical plasmoids forming vortex loops around a ring, similar to Parson's toroidal ring model. Parson's model was further elaborated by a group of American physicists that included Charles Lucas, Jr., Glen Collins, and David Bergman. In 2000 two American theoretical physicists, Jean-Luc Cambier and David Micheletti, introduced the toroidal spiral concept of an electron based on the cold-fluid equations for plasma shell coupled with the self-generated magnetic field.

Louis de Broglie (1892-1987) – In 1923, while working on his Ph.D. thesis, the French physicist Louis de Broglie asked himself a logical question: if waves behave like particles, must not the reverse also be true, that particles behave like waves? De Broglie generalized this statement by making a profound proposition that the motion of each particle is accompanied by a wave, which guides the particle through space. Moreover, it was not an electromagnetic wave but a mechanical wave, propagating faster than the velocity of light. He then found that the wavelength λ of this wave is related to the Planck constant h, particle mass m, particle velocity V and particle momentum p by a simple equation:

$$\lambda = \frac{h}{mV} = \frac{h}{p}$$

Now de Broglie was ready to apply his theory to the electron in Bohr's model of the hydrogen atom. His goal was to explain that only certain electron orbits are allowed. Equipped with his idea of the wave accompanying an electron as it moves around a nucleus, de Broglie envisioned this wave as a standing wave similar to a musical sound wave radiating from a violin. In the violin, the standing waves are produced by interaction between its strings and body. The strings of a violin are stretched across the bridge and nut so the string ends become stationary.

De Broglie proposed that all of the possible states of the electron in the atom are standing wave states, each with its own wavelength (Fig. 1.5.2). The electron in the atom no longer appears as a planet in a solar system, but more like a vibrating circular string.

By using his model, de Broglie provided a visualizable explanation of the allowed states of the electron orbits. One simply needs to assume that the allowed states arise because the electron matter waves form standing waves, when an integral number n of wavelengths fits exactly into the circumference of a circular orbit. Thus, for the orbit with radius r we obtain that $nh = 2\pi r$. This relationship leads precisely to Bohr's equation for quantization of electron orbits that we described earlier.

Erwin Schrödinger (1887-1961) - Soon after de Broglie published his paper in 1925, the Austrian physicist Erwin Schrödinger dealt another devastating blow to the particle nature of the electron. He dared to claim that electrons are not particles at all, but classical waves, like ordinary sound waves, water waves or electromagnetic waves. Moreover, the particle aspect of electrons is merely illusory.

Schrödinger came up with a differential equation in which the electrons were presented as matter waves. The principal parameter in his equation was the so-called *wave function* ψ (Greek: psi) that is dependent on distance and time. Schrödinger solved his equation in application to atomic electrons, producing the wave functions ψ corresponding to all possible orbits of the electrons. In Schrödinger's interpretation, the square of the amplitude of wave function ψ^2 described the density distribution of matter waves. Not everyone agreed with the wave theory of electron.

Among the opponents of Schrödinger's wave theory was the German physicist Max Born, who insisted that the electron is still a particle and only the wavy part, still related to the electron, is Schrödinger's wave function. But this function has nothing to do with the electron structure. Moreover, the square of the amplitude of the wave function ψ^2 is not the density distribution of matter waves, as Schrödinger believed, but is the probability of finding an electron at various locations in a certain quantum state. So, according to Born, the wave function is a *wave of probability*. The opinions of scientists about the validity of Born's interpretation were not unanimous. Bohr, Heisenberg and Sommerfeld had no issues with Born's idea, but Schrödinger, de Broglie and Einstein found it totally unacceptable. The dramatic split of opinion between these outstanding physicists eventually led to several bitter but productive confrontations over the years following.

Werner Heisenberg (1901-1976) –Amidst the confusion created by the introduction of the idea of wave probability, the German theoretical physicist Werner Heisenberg introduced, in 1927, another even more controversial concept that became known as *Heisenberg's uncertainty principle*. The logical basis for the development of this concept was quite simple. During an experiment we are using sensors to measure certain parameters of an object under investigation. For instance, if we want to measure the water temperature in a swimming pool, we use a temperature sensor that we place into water. We are not concerned if, before we took the measurement, the sensor was in a shady area or under a hot sun, and rightfully so. The amount of heat contained inside the sensor is miniscule compared to the heat contained in the water of the swimming pool. Therefore, the temperature of the sensor has no appreciable effect on the accuracy of our measurement.

But it is a completely different matter if we decide to measure the temperature of water contained in a very small glass tube. In that case, the amount of heat contained in the sensor may become sufficient to affect the temperature of the water in the glass tube, distorting the results of our measurement. A somewhat similar "thought experiment" was probably in Heisenberg's mind when he pointed out that a similar situation occurs when we do our measurements on atomic scale. Let us suppose that we want to measure a position and the velocity of an atomic electron located at a certain quantum energy state. We will use, of course, the tiniest sensor available to us that utilizes a beam of light. But, as we know from the photoelectric effect, the photons will excite the electron. Therefore, its trajectory will be distorted during the process of measurement. Based on this logic, Heisenberg concluded that it is fundamentally impossible to make simultaneous measurement of a particle's position and velocity (or momentum) with infinite accuracy. He stated in his famous principle:

If a measurement of position is made with precision Δx and a simultaneous measurement of momentum is made with precision Δp, then the product of the two uncertainties can never be smaller than a number of the order of $h/(2\pi)$; That is,

$$\Delta x \cdot \Delta p \geq \frac{h}{2\pi}$$

Heisenberg's uncertainty principle became an indispensable addition to quantum theory. But, as with Born's idea of probability waves, not everyone accepted Heisenberg's uncertainty principle. The confrontation between the two different theories of physics stimulated the thinking of many young physicists, and in some cases led to a better understanding of the "spooky" world of quantum physics.

Roger Penrose (b. 1931) - The British theoretical physicist Roger Penrose became well known for his contributions to theories of black holes, but his most favorite idea was certainly the *Twistor theory* that he conceived in mid-1960s. At the core of his theory was a mathematical entity called the *twistor* that was capable of defining empty space, all the subatomic particles and the four forces, including gravity.

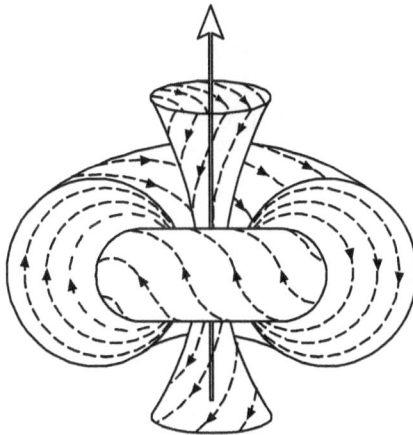

Figure 1.5.2. The Penrose's twistor. Adapted from
J. Boslough (1992).

The twistor was supposed to be at nature's most fundamental level. To accomplish this task, Penrose thought, the twistor must combine both translational and rotational motions; that is, it must be both spinning and moving along. In addition, it must be both quantum and relativistic. Penrose envisioned the twistor as a pair of concentric doughnuts with a short shank of rope through its hole (Fig. 1.5.2). The size of the twistor was somewhere between an absolute point and the size of an elementary particle. By combining various types of twistors it was possible to produce certain elementary particles. But massless particles, such as photons or neutrinos, could be made from a single twistor.

Similarly to the general theory of relativity, Penrose's twistor theory held that gravity was caused by mass giving rise to a curvature of the universe. To solve the problem mathematically, Penrose employed the theory of complex numbers and other advanced theories of mathematics. Unfortunately, the equations turned out to be very complicated, even for Penrose, one of the world's foremost mathematicians.

Richard Gauthier (b. 1946) - Since 1993, the American theoretical physicist Richard Gauthier has been exploring the hypothesis that matter and energy can be composed of helically moving point-like quantum particles that may move faster than light. According to his hypothesis that he described below, these particles are called transluminal energy quanta, or TEQs. They may move along open helical paths and form photons - quantum particles of visible light or other electro-magnetic energy such as radio waves, x-rays and gamma rays. Figure 1.5.3 shows transluminal energy quantum (TEQ) models of three photons of different wavelengths. The TEQs are the three black dots moving to the right along open helical trajectories at a velocity 1.414 (the square root of 2) times the velocity of light.

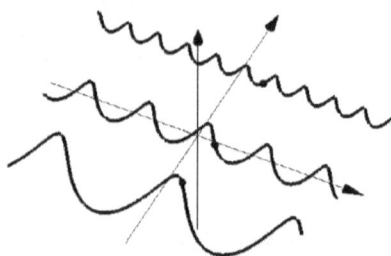

Figure 1.5.3.
Transluminal energy
quantum (TEQ) models of
three photons of different
wavelengths.

Alternatively, they may move in closed helical paths to produce various particles having mass, such as the electron, the muon and the tau particle (the family of electron-like particles). A TEQ carries energy, momentum and spin and sometimes charge (as with the TEQ electron model.)

When a TEQ moves in a closed helical path, it moves along the mathematical surface of a torus. A TEQ can moves at velocities slower and faster than velocity of light, even passing through the velocity of light many times per second when composing a particle with mass like the electron.

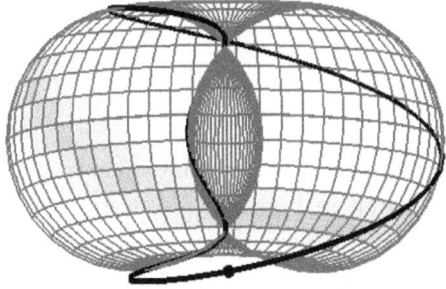

Figure 1.5.4.
Transluminal energy quantum model of the electron.

As shown in Figure 1.5.4, the TEQ moves from a minimum velocity of 0.707 times the velocity of light to a maximum velocity of 2.516 times the velocity of light, passing frequently through the velocity of light. The self-intersecting mathematical toroidal surface on which the TEQ moves is partly cut away to show the complete TEQ trajectory.

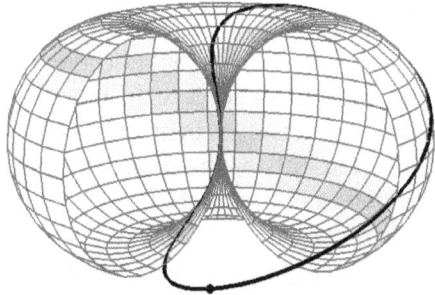

Figure 1.5.5.
Transluminal energy quantum model of the cosmic quantum.

In addition to the photon and the electron, two hypothetical particles have been modeled with the TEQ. The first is the cosmic quantum (Fig. 1.5.5), which is the hypothetical first particle of the universe. The cosmic quantum would have contained all the positive initial energy of the universe in a single tiny quantum particle much smaller than a single atom. It is a spin-1 particle or boson.

Here the TEQ's velocity varies from the velocity of light (at the minimum) to 2.236 times the velocity of light (at the maximum).

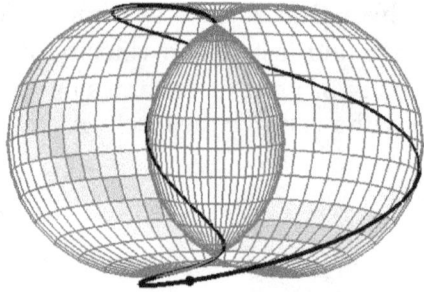

Figure 1.5.6.
Transluminal energy quantum model of a dark matter.

The second hypothetical particle is a particle of dark matter. As shown in Figure 1.5.6, it is a spin-$\frac{1}{2}$ particle or fermion. Here the TEQ's minimum velocity is the velocity of light and its maximum velocity is 3.162 times the velocity of light. Dark matter is considered to compose about 83% of all the matter in the universe. Most of the dark matter in the universe is generally thought to be composed of non-baryonic elementary particles that mainly react with other particles through gravitational attraction.

1.6 The Standard Model

There was an explosion in the discovery of new elementary particles in the second half of the 20th century, thanks to the construction of powerful particle accelerators. There was a need to find some kind of order inside the ever-expanding "zoo of particles." Many physicists rightfully considered Mendeleyev's Periodic Table of Elements as the model to follow. By the end of the 1970s the knowledge accumulated by particle physicists evolved into the so-called *Standard Model* that provides the description of the world in terms of truly fundamental particles, *leptons* and *quarks*. All other particles, the hadrons, are merely composites of the leptons and the quarks. The word "lepton" came from the Greek word *leptos* meaning "small or light." The name "hadron" came from the Greek word *hadros*, which means "thick and heavy." The hadrons are divided into two classes, called *mesons* and *baryons*. The name "meson" originated from the Greek word *mesos*, which means "in between." The name "baryon" derived from the Greek word *barys* for "heavy." Proton is the lightest baryon.

Based on their principal function, the particles are divided into two groups: the mass-carrying particles called *fermions* and the force-carrying particles responsible for interaction between particles called *bosons*. Standard Model considers five principal forces operating between particles: color, weak, electromagnetic, gravitational and Higgs.

- <u>Color force</u> operates between quarks, and also shows up as the strong force acting between nucleons. The color force is mediated by *gluons*.
- <u>Electromagnetic force</u> operates between electrically charged particles. It is mediated by *photons*.
- <u>Weak force</u> is responsible for particle decay. It is mediated by the *intermediate vector bosons* W^+, W^- and Z^0.
- <u>Gravitational force</u> operates between all particles. It is mediated by *gravitons*.
- <u>Higgs force</u> is responsible for the creation of the masses of all particles. It is mediated by the *Higgs boson*.

Quark model – The quark model is a cornerstone of the Standard Model. The idea of quarks was proposed independently by the American theoretical physicists Murray Gell-Mann (b. 1929) and George Zweig (b. 1937). According to this model, all elementary particles, except for leptons, are made of fundamental particles called *quarks* which have fractional charges. Although Zweig and Gell-Mann were talking about essentially the same physics, they treated quarks in completely different ways. Gell-Mann thought of a quark as a "mathematical" entity helping to find some order among subatomic particles. To Zweig, quarks were real, concrete particles, providing a real challenge for experimenters to find them.

In the original quark model, the quarks came in three *flavors*: u (up), d (down), and s (strange), but by 1977, the model was expanded by several physicists to include three additional flavors: c (charmed), b (bottom) and t (top). The u, d, s, c, b, and t quarks have respective fractional charges Q equal to $+\frac{2}{3}$, $-\frac{1}{3}$, $-\frac{1}{3}$, $+\frac{2}{3}$, $-\frac{1}{3}$, and $+\frac{2}{3}$. The antiquarks have opposite electric charges. All quarks have the same spin $J = \frac{1}{2}$ and the same baryon number $N = \frac{1}{3}$. Additionally, there are four quantum numbers assigned to each quark, strangeness S, charm C, bottomness B, and topness T.

Notably, except for the fractional charges Q and the spin J, the other five quantum numbers did not have any physical meaning.

Formation of hadrons - It is easy to use the quark model for defining the compositions of hadrons. The main feature of hadrons is their ability to react to the strong force. The baryons (including proton and neutron) are identified by the baryon number N that is equal to +1 for the particles and -1 for the antiparticles. For all particles other than the baryons, $N = 0$.

One should follow three simple rules when using the quark model (Fig. 1.6.1):

1. Each meson is made of one quark and one antiquark.
2. Each baryon is made of three quarks.
3. Each antibaryon is made of three antiquarks.

Leptons include three particles and three antiparticles: electron e, muon μ and tau τ. There is either a neutrino or an antineutrino associated with each lepton. Mesons are made up of two quarks, while baryons are made up of three quarks.

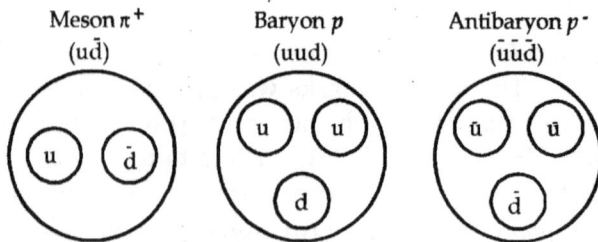

Figure 1.6.1. Quark compositions of several particles.

Since the discovery of the first meson in 1937, the number of discovered particles that belongs to the meson class had been increasing at an exponential rate. By the end of the 20th century this class included more than 340 meson particles and antiparticles, and some of them were more than ten times heavier than protons. So, the original meaning of meson as something "in between" became obsolete. To sort out all these particles, it was necessary to split the meson class further into several subclasses.

Notably, among discovered particles only the electron, the positron and the proton are stable, in other words, they never disintegrate into something else. Photons and neutrinos are also believed to be stable. All other particles are not stable, including the free neutron, the average lifetime of which is about 15 minutes.

Higgs mechanism - The quark model has a serious problem with the quark masses. The total mass of the quarks making up an elementary particle is much smaller than the actual mass of that particle. To solve this problem, the British cosmologist and particle physicist Peter Higgs (b.1929) proposed that there is an undetectable field that is now called the *Higgs field*. The Higgs field is associated with a boson that has a big mass. This particle became known as the *Higgs particle*, or the *Higgs boson*. The Higgs mechanism involves swallowing up the Higgs boson by any photon-like particle, providing the particle with mass. According to different sources, the predicted mass of the Higgs boson varies from 80 to 1000 GeV. The collisions of protons at the CERN Large Hadron Collider in 2012 produced a particle with the mass ranging from 125 GeV to 126 GeV. Some physicists believe that this is the expected Higgs boson. The results of these experiments could be confirmed more conclusively in 2013 when the power of the Large Hadron Collider will be doubled.

Interactions Between Charged Light Particles – The interactions between charged light particles, such as electrons, are described by the *theory of quantum electrodynamics*, or *QED*, that is a part of the Standard Model. Three physicists are mostly credited by the scientific community for the development of QED: the two American theoretical physicists Richard Feynman (1918-1988) and Julian Schwinger (1918-1994), and the Japanese theoretical physicist Sin-Itiro Tomonaga (1906-1979). According to QED, the interactions between charged light particles like electrons are mediated by photons. QED uses very complicated equations of quantum mechanics to describe the interactions between particles.

Feynman introduced his famous *Feynman diagrams* that helped many thousands of particle physicists around the world to describe very sophisticated nuclear reactions by visualizing particle collisions and talking about them in a common language. Consider, for instance, Feynman's spacetime diagram depicting the simplest collision of two electrons, shown in Fig. 1.6.2. This is a typical case of an *electromagnetic interaction*. The straight lines represent the paths of the electrons e- and the wiggly line the path of the virtual photon γ. The virtual photon fires up when the distance between the electrons reaches a certain minimum distance. Consequently, the mediation of the virtual photon produces the electromagnetic interaction that pushes the colliding electrons apart. Thanks to the Feynman diagram, the process of collision of two electrons appears very simple, but a particle physicist knows that behind each line in the diagram hides a specific term in the complex mathematical expression for the probability of this collision.

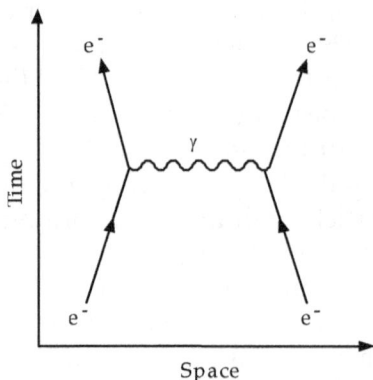

Figure 1.6.2.
Feynmann diagram depicting the collision of two electrons.

Feynman, Schwinger and Sin-Itiro Tomonaga were involved in attacking the principal problem of QED known as the *War Against Infinities*. According to QED, interactions between charged light particles such as electrons are described as being mediated by photons. The problem arises when one wants to take into account the so-called *self-interaction* between the electron and its own electric field. Based on the laws of interaction between the electric charge and electric field, the mass of the electron must be infinite.

Feynman solved the problem with the aid of a mathematical trick known as *renormalization*. Renormalization was introduced in its complete form in 1947 by the Dutch physicist Hendrik Kramer (1894-1952). It was based on his assumption that the mass of an atomic electron is made of two components: the mass measured at very short range, *bare mass*, and the *infinite mass*, produced by electromagnetic self-interaction. Bare mass can be defined by subtracting the infinite bare mass from the infinite self-interaction mass of the electron. If renormalization of a proposed theory produced correct results, then the physicists considered the theory *renormalizable*, and, thus, acceptable. Eventually, the capability to be renormalized became one of the most important criteria for validating a physical model. Nevertheless, this technique still bothered many physicists, including Paul Dirac. Feynman himself said in his book *QED* that renormalization "is what I would call a dippy process!"

Weak & Electroweak Interactions Between Particles - The weak force, like the other forces in the subatomic world, operates by the exchange of messenger particles, but with one principal difference. Instead of pushing particles apart or pulling them together, the exchange of the messenger particles modifies the character of the particles that swap them. In the weak interactions, the messenger particles are the charged bosons W^+ and W^-. For instance, during beta decay a neutron n emits an electron e^- and an antineutrino $\bar{v}_e$, converting itself into a proton p according to the equation:

$$n \rightarrow p + e^- + \bar{v}_e$$

The above equation, however, does not provide us with a full description of the weak interaction during the beta decay. Here again, the Feynman diagram becomes very helpful. As shown in Figure 1.6.3, during beta decay, the down quark d inside a neutron interacts with an incoming electron antineutrino $\bar{v}_e$ through a messenger particle W^-. Consequently, the down quark d converts into the up quark u, while the electron antineutrino $\bar{v}_e$ turns itself into the electron e^-.

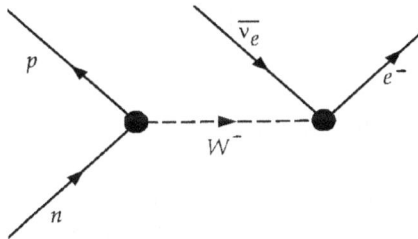

Figure 1.6.3. Feynman diagram depicting neutron decay.

In 1960, Sheldon Glashow proposed that the neutral weak current was carried by a massive neutral particle Z^0. This particle would mediate, for instance, the exchange force during the collision between the gauge bosons W^+ and W^-. The old problem with infinities, however, immediately showed up in Glashow's theory. The solution was proposed by another American physicist, Steven Weinberg (b. 1933), who first pointed out on the possibility of getting rid of the infinities in weak interactions by employing the Higgs mechanism. Eventually, jointly with the Pakistani theoretical physicist Abdus Salam (1926-1996) and the Dutch theoretical physicist Gerald't Hooft (b. 1946), Sheldon Glashow and Steven Weinberg proved that by using the Higgs mechanism their theory becomes renormalizable.

Strong Interactions Between Particles - The strong interactions occur between nucleons. It holds nucleons together within the nuclei. The quark model worked very well for the construction of most subatomic particles, but there were serious difficulties with making several particles. The solution of this problem was as always swift: simply propose a new quark to possess one more property. The physicists with artistic minds named this property *color*, and the theory of interaction between quarks became known as *quantum chromodynamics*, or QCD.

In QCD, everything comes in color that occurs in three varieties: *red, green* and *blue*. Each quark carries a *color charge*, in analogy to electric charges. For each quark of a certain color there is a quark with the opposite color. In another analogy with the electric force between electric charges, the force between quarks is called the *color force*. The color force is thought to be mediated by massless particles called *gluons*, in analogy with the photons responsible for mediation of electric force. Notably, in QCD we are dealing with the dynamic quarks that continuously mingle with one another by exchanging their colors.

Notably, there is no physical meaning behind the "color." The color is employed merely to allow for more varieties of elementary particles. Instead of different colors one could use, for instance, different kinds of fruits or vegetables.

Symmetry - Intuitively, physicists believed that our world must be symmetrical. In application to the subatomic world they considered several spacetime symmetries that must be conserved during the interactions between particles. The most important among them were related to the conservation of *parity* (P), *charge conjugation* (C), and *time-reversal invariance* (T).

Parity (P) is the operation which reverses the signs of coordinates to describe a system. Thus, a position described in three dimensions by the coordinates x, y, and z is now described as the positions $-x$, $-y$, and $-z$. This is equivalent to studying the mirror image of the original system. For instance, a mirror image of a particle with a certain spin will produce a particle with the opposite spin.

Charge conjugation symmetry (C) states that every subatomic particle has an antiparticle with the same mass and opposite values of all other quantum numbers, such as charge. In respect to charge this symmetry appears to be the same as (P) symmetry, but (C) symmetry is more general because it requires other parameters, besides charge, to be symmetrical. Therefore, it can also be applied to neutral particles.

Time-reversal invariance symmetry (T) states that the properties of particles will not change if the direction of time changes. The (P), (C) and (T) symmetries turned out to be more elusive than the physicists had expected. Although parity is conserved in electromagnetic interactions, it refuses to be maintained for the weak nuclear interactions. This odd behavior of the weak nuclear interactions got the name *broken symmetry*.

1.7 Unified Theories of Physics

As we described before, the first serious attempts to develop a unified theory of physics were made in the beginning of the 20th century by the German mathematician Hermann Weyl and the Swedish theoretical physicist Oskar Klein. Albert Einstein accepted neither Weyl's nor Kaluza-Klein's theory. After numerous unsuccessful attempts to modify them to his satisfaction, he eventually decided to abandon both of them. But, in spite of their demise, these theories certainly stimulated Einstein's efforts to develop a unified field theory, a monumental effort that lasted for more than three decades.

It is well known that Einstein did not succeed. This failure, however, did not discourage many young physicists from following in his footsteps. But in fact, their task was even more difficult. During Einstein's time there were only two kinds of known forces, electromagnetic and gravitational. The new generation of physicists had to deal with three additional forces, color, weak, and Higgs. But the challenge was clear and, in mid-1970s, when the Standard Model was at the peak of its success, it seemed that it was only a matter of time before the quantum mechanics of the Standard Model would be unified with the curved spacetime gravity of the general theory of relativity.

The unified theories of physics developed during the 20th century are divided into two groups, the Grand Unified Theories (GUTs) and the Theories of Everything (TOEs). GUT is a gauge theory that provides the same mathematical tool to describe the electromagnetic, the weak nuclear and the strong nuclear forces. The Standard Model is a typical representative of GUT. The most popular alternative to GUT is the Supersymmetry Theory that carries the cute name *SUSY*. The first realistic version of this theory was proposed in 1981 by the American physicist Howard Georgi (b. 1947) and the Greek physicist Savas Dimopoulos (b. 1952). This theory gets rid of the last major asymmetry in the world of sub-atomic particles that divides them into fermions (the particles that make matter) and bosons (the particles that carry force).

To accomplish this unification task, SUSY provides a boson partner for every fermion and a fermion partner for every boson. This happy marriage of particles is accompanied by an easy way of naming the partners. For instance, the SUSY partners of electrons and quarks are respectively called *selectrons* and *squarks*, while the SUSY partners of photons and gluons are respectively called *photinos* and *gluinos*. But all this happiness has its price. The downside of SUSY is two-fold. Firstly, it doubles the already excessive number of particles that are present in the Standard Model. Secondly, it works only by adding another four dimensions to the four dimensions of ordinary spacetime. Another GUT, called SU(5), was proposed in 1974 by Howard Georgi and Sheldon Glashow. It includes all forces except gravity. The theory makes a prediction, still unproven, that protons could decay.

The end of the 20th century appeared to be a fair time mark to complete the task. It did not happen. Instead, some scientists have begun abandoning the Standard Model as the basis for a unified theory of physics.

1.8 String, Supergravity & M-Theories

The origin of the string theory goes back to 1968. That year the Italian theoretical physicist Gabriele Veneziano (b. 1942) noticed that a two-hundred-year-old formula, known as the *Euler beta function*, seemed to match the data on the strong force with surprising precision. Was it a pure coincidence? In 1970, the American theoretical physicist Leonard Susskind (b. 1940), the Danish theoretical physicist Holger Nielsen (b. 1942) and the American physicist Yoichiro Nambu (b. 1921) recognized the physical meaning of Veneziano's discovery. They showed that if two particles are connected by a tiny, extremely thin rubber-like string, then the strong forces between these particles are described by Euler's equation.

In 1974, string theory was further extended by one of the earliest string enthusiasts, the American theoretical physicist John Schwarz (b. 1941) and his collaborator, the French theoretical physicist Joël Scherk (1946-1980). The two physicists proposed a string theory of quantum gravity. But it took ten years for this theory to be widely accepted. String theory treats particles in a very special way. According to a conventional theory, such as the

Standard Model, the electrons and the quarks are simply the dots with no spatial extent in any direction. This is not so according to string theory: it presents the particles as tiny, vibrating filaments of energy, called *strings*.

The strings have no dimensions except for length, making strings one-dimensional entities. At the same time, because the strings are extremely small (about 10^{-35} m), they appear to be points whenever they are detected by the most sensitive devices in the world. The string vibrates similarly to a string of a violin or a cello. But unlike music, where we are mostly concerned with different tones, string theory deals with different vibrational patterns of the strings (Fig. 1.8.1) that make up both the matter particles (fermions) and the force carrier particle (bosons).

Each specific vibrational pattern corresponds to certain values of the particle mass, electric charge, and spin. In addition, the American theoretical physicist John Schwarz (b.1941) and the French theoretical physicist Joël Scherk (1946-1980) showed in 1974 that there is an additional vibrational pattern that has all the properties of a graviton. Thus, string theory included gravity. One of the serious drawbacks of string theory is that it requires many dimensions. The original version, called *bosonic*, required 26 dimensions, while, in the later versions, this number was reduced to ten. The reduction of the dimensions became possible by applying Kaluza's compactification described before. Another difficulty with the original version was that it required a particle called *tachyon* that travels faster than light.

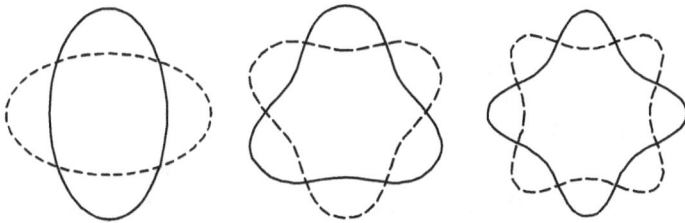

Figure 1.8.1. Some examples of the string vibrational patterns. Adapted from B. Greene (2004).

As we know, according to Einstein's special theory of relativity, the velocity of light is a kind of barrier no material body can trespass. However, superluminal velocity is not forbidden by the equations of his theory for the particles already existing on the other side of the light barrier. Einstein's equations predict that the mass of tachyons must be expressed by imaginary numbers, and these particles can never be slowed down to reach the velocity of light. An introduction of the tachyon added another kind of symmetry to the world of particles that now could travel both below and above the velocity of light. The string theorists, however, preferred to abandon this symmetry rather than to deal with a controversial superluminal velocity. The tachyons did not show up any more in the later versions of superstring theories.

The next significant improvement of the string theory came in 1980, when several physicists discovered that patterns of string vibrations come in pairs (super-partner pairs), differing from each other by a half unit of spin. The new version of the string theory that incorporates this novel symmetry was named the *superstring theory*. Still, the new theory was plagued with deadly quantum anomalies that threatened to make it senseless. But in 1984 the American theoretical physicists Michael Green (b. 1946) and John Schwarz successfully demonstrated that the superstring theory would be free from quantum anomalies if two conditions are met. Firstly, the theory must employ ten spacetime dimensions and, secondly, it must use a special type of the quantum gauge symmetry group known as SO(32) or E(8)×E(8).

This breakthrough helped to bring attention of many particle physicists to this theory. The triumphant march of the superstring theory continued and, in 1984, four physicists at Princeton – David Gross, Jeffrey Harvey, Emil Martinec, and Ryan Rohm (dubbed the *Princeton String Quartet*) discovered another class of anomaly-free superstring theory with the quantum gauge symmetry group E(8)×E(8). Their version of the superstring theory had even better properties than the Green-Schwartz version. By the end of the 20th century, three additional versions of the superstring theory showed up along with another promising theory called the *supergravity*. Can be the string theory unified with the theory of supergravity? Some physicists believe that the supergravity and string theory are just parts of the so-called *Membrane*, or *M-theory*, in which the strings are replaced with membranes. Only time will tell if this proposition has some merits.

1.9 The Space & Information Age

As the skepticism about the bright prospects of both the Standard Model and superstring theory grew stronger, some of the most daring scientists began questioning the basic assumptions of these theories. Some scientists also started reviewing various aspects of Einstein's theory of relativity. Particularly, they went as far as to reexamine the most fundamental postulate of the theory, that of the maximum velocity in the universe.

The EPR Paradox - Paradoxically, Einstein, the man proclaiming that the velocity of light is the maximum velocity in the universe, initiated a debate that led to the discovery of instantaneous transmission of information between particles. Einstein was always bothered by Heisenberg's uncertainty principle. Remember his famous saying that God does not play with dice? So, he was periodically creating new "thought experiments" in his attempts to disprove Heisenberg's concept, each time to be defeated by his clever opponent, Niels Bohr.

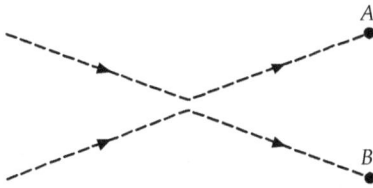

Figure 1.9.1.
Interaction
between
two electrons.

In 1935 he, with the help of his two collaborators, the American physicist Boris Podolsky (1896-1966) and the American-Israeli physicist Nathan Rosen (1909-1995), came up with the idea of the eponymous *EPR experiment*. The essence of their thought experiment is deceptively simple. Imagine two electrons A and B that, upon colliding, fly a long distance apart (Fig. 1.9.1). Suppose that we were able to measure the total angular momentum of the system that includes both electrons at the time when they interacted. According to the law of the conservation of momentum, the total momentum of two electrons cannot change as long as they do not interact with anything else after their collision. Thus, if we were to measure, much later, the momentum of the electron A, then it would be possible to calculate the momentum of the electron B by subtracting the measured momentum of the electron A from the total.

The same is true if, instead of measuring the momentum of the electron A, we measure its position. According to quantum theory, if we observe the precise position of electron A, we can, at the same moment, calculate the precise position of electron B. Heisenberg's uncertainty principle tells us that we cannot measure precisely both the momentum and the position of the electron. So, if this rule is strictly maintained, then whenever we measure the precise position of the electron *A*, the electron *B* will also have a precise position, but both electrons will have uncertain momentum. Conversely, whenever we measure the precise momentum of the electron *A*, the electron *B* will also have a precise momentum, but the positions of both electrons will be uncertain. Einstein argued that this contradiction can be solved only by accepting Newton's "spooky action at a distance," according to which the information from one electron to another propagates instantaneously. But he could not accept this solution because it contradicted the most fundamental postulate of his special theory of relativity - that nothing can move faster than the velocity of light.

In 1952, three years before Einstein's death, the American physicist David Bohm (1917-1992) came up with his version of the EPR experiment. Although it was originally designed by Bohm as a thought experiment, the advances in the technologies of measurement have made it possible to conduct a real experiment. In 1964 the Irish physicist John Bell (1928-1991) demonstrated that Bohm's thought experiment could be practically implemented by measuring the polarization of photons. Crucial to the success of his method was the so-called *Bell's inequality* that provides an objective criterion for the treatment of the results of the experiments.

Several experiments were conducted by following the ideas of Bohm and Bell. The most prominent of them were carried out in Paris in the early 1980s by the French physicist Alain Aspect (b. 1947) and his colleagues. After the publication of their paper in 1982, the idea of action at a distance, no longer spooky, took its legitimate place in the world of quantum mechanics. Lately, a different term, *non-locality*, became more popular in describing the same concept. Non-locality underlines that the behavior of quantum entities is affected not only by what is going on at the point where the entity is located, but also by events that are occurring at other localities, which may in principle be far away, even across the Universe. So, one may rightfully conclude that Einstein turned out to be wrong. The events in the subatomic world can be correlated instantaneously. But at the same time, thanks to his inquisitive mind, by proposing a challenging paradox, he stimulated the thinking of many bright scientists who discovered one of the most puzzling phenomena in the world of quantum physics.

Reversal Of Time - The idea of instantaneous transmission of information between subatomic particles leads to another

"spooky" concept, that of *negative time,* when time goes back-wards. Let us start from Maxwell's theory of electromagnetism.
Initially, its differential equations were thought to be applicable only for positive, or conventional, time.   But between 1929 and 1932 the Dutch physicist Adriaan Fokker (1887-1972) proved that Maxwell's equations are completely symmetrical in time. Simply saying, whether you will input in these equations the time vari-able −*t* or *t,* the result of your calculations will be exactly the same.
In 1940 Richard Feynman and John Wheeler (1911-2008) applied this concept to their *absorber theory.* The principal assumption of this theory is that an excited electron emits two types of waves, the *retarded waves* that travel forward in time, and the *advanced waves* that travel backwards in time.

Consider two electrons separated by a certain distance (Fig. 1.9.2). After its excitation, the first electron emits simultaneously two half waves, a half retarded wave *R1* and a half advanced wave *A1* that propagate in the opposite directions with the veloc-ity of light. After the half retarded wave *R1* reaches the second electron, this electron will emit in response a half retarded wave *R2* and a half advanced wave *A2* that also propagate in the oppo-site directions with the velocity of light.

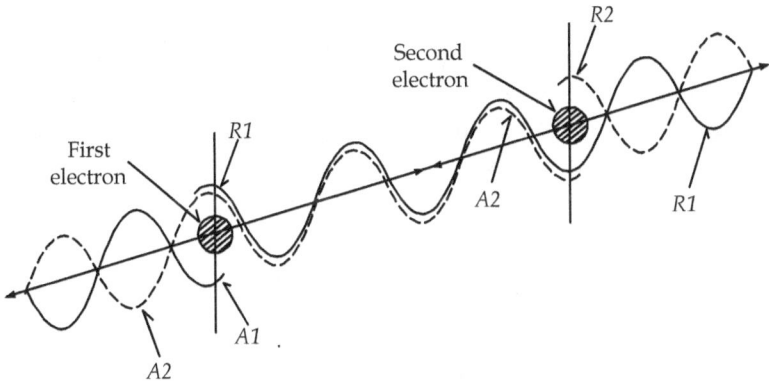

Figure 1.9.2. Instantaneous transmission of light according to Wheeler-Feynman absorber theory. Adapted from J. Gribbin (2000).

The two sets of waves cancel out everywhere except in the re-gion between the two electrons. Since within this region the half retarded wave *R1* travels forward in time and the half advanced wave *A2* travels backwards in time, the connection between the two electrons is made simultaneously.

Quantum vacuum - To explain some bizarre properties of field and matter, quantum mechanics required the existence of the so-called *quantum vacuum,* or *zero-point energy.* The origin of the zero-point energy is closely associated with the name of an outstanding German physical chemist Walther Nernst (1864-1941). In 1916 he proposed that empty space was filled with zero-point energy. Eventually, zero-point energy was defined as the energy which is still associated with a particle or a system (over and above its mass-energy) at the absolute zero of temperature, 0^0 Kelvin. Still, the origin of zero-point energy was not clear. In 1927, after Heisenberg introduced his uncertainty principle, physicists agreed that this principle required the existence of zero-point energy. Belgian astronomer George Lemaître (1894-1966), the originator of the *Big Bang theory* for the origin of the universe, believed that vacuum energy played an important role in the creation of the universe.

In quantum theory, the vacuum ground state is not completely empty, but contains a seething mass of virtual particles and fields. One can compare these fields with ripples of water in the ocean. This may explain why these fields called *vacuum fluctuations.* In 1948, the Dutch physicist Hendrik Casimir (1909-2000) predicted that the electromagnetic zero-point energy fluctuations of the vacuum can be observable on microscopic objects. He noted that the interactions between two neutral molecules can be interpreted in terms of quantum vacuum fluctuations that produce virtual photons.

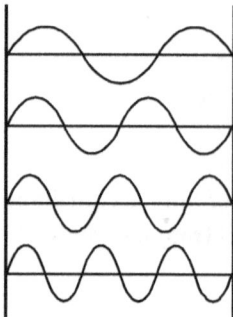

Figure 1.9.3.
Standing waves
between
conducting plates.

Casimir, as a thought experiment, replaced two molecules with two metallic plates with their highly reflective surfaces facing each other in a vacuum. He then envisioned the virtual photons bouncing between the plates. Since the photons behave like electromagnetic waves, the electromagnetic field between the plates would be amplified, if integer multiples of half a wavelength could fit exactly between the plates (Fig. 1.9.3). This phenomenon is now known as the *Casimir effect*.

Thus, the pressure from the photons, as they bounce between the plates, will depend on the distance between the plates. But outside of the plates all of the frequencies are of equal importance. Therefore, the pressure from the photons bouncing off the outer surfaces of the plates will not depend on the distance between plates. Casimir showed that the total pressure P (or force F per unit plate area A) from the inside and the outside of the plates will be attractive, trying to bring the plates closer to each other. This pressure is inversely proportional to the fourth power of the distance a between the plates:

$$P = \frac{F}{A} = \frac{\pi}{480} \frac{hc}{a^4}$$

The notion of the vacuum space filled with radiation and vacuum fluctuations eventually led some physicists and astrophysicists to propose several hypotheses in which the entire universe is merely a particular form of vacuum. The total energy of the Universe may balance out to zero, while locally there may be positive and negative energy regions. According to one of these hypotheses, proposed by the American physicist Edward Tryon in 1973, the universe may be a large-scale quantum mechanical vacuum fluctuation, where positive mass-energy is balanced by negative gravitational potential energy.

Table 1.9. Regions of the electromagnetic radiation.

Region	Wavelength cm	Frequency Hz
Radio waves	> 10	$< 3 \times 10^9$
Microwave	$10 - 10^{-2}$	$3 \times 10^9 - 3 \times 10^{12}$
Infrared	$10^{-2} - 7 \times 10^{-5}$	$3 \times 10^{12} - 4.3 \times 10^{14}$
Visible	$7 \times 10^{-5} - 4 \times 10^{-5}$	$4.3 \times 10^{14} - 7.5 \times 10^{14}$
Ultraviolet	$4 \times 10^{-5} - 10^{-7}$	$7.5 \times 10^{14} - 3 \times 10^{17}$
X-rays	$10^{-6} - 10^{-9}$	$3 \times 10^{16} - 3 \times 10^{21}$
Gamma rays	$< 10^{-9}$	$3 \times 10^{18} - 3.4 \times 10^{22}$

Cosmic Messengers - The 20th century was a turning point in astronomy. New technology enabled astronomers to detect cosmic radiation covering a very wide range of wavelengths (Table 1.9).

Radio waves - Radio waves have the longest wavelengths in the electromagnetic spectrum. Radio wavelengths vary widely, ranging from ten centimeters to several hundred meters. Incidentally, AM radio uses the longest wavelengths, while FM radio uses the shortest ones. The main advantage of radio waves over visible light is their capability to penetrate cosmic dust, so one can see not only the galactic center but also the opposite side of the galaxy.

Microwave radiation – We know that microwaves are used for heating food. They are also good for transmitting information, because they can penetrate haze, light rain and snow, clouds and smoke. One of the most popular cosmological theories of the 20th century is closely related to microwaves. In 1948, the Russian-born American physicist Georgy Gamow (1904-1968) and the American cosmologist Ralph Alpher (1921-2007) published their version of the Big Bang theory. They predicted that, as a result of a cataclysmic explosion, there must be a remnant radiation observable in the microwave region of the spectrum. Seventeen years later, in 1965, two American radio astronomers, Arno Penzias (b. 1933) and Robert Wilson (b. 1936), discovered this radiation, which became known as *cosmic microwave background* (CMB) radiation.

Infrared radiation – Infrared light lies between the visible and microwave portions of the electromagnetic radiation spectrum. They are divided into two sub-ranges, near infrared and far infrared. The wavelength of near infrared is closest to visible light, while the wavelength of far infrared is closer to the microwave region of the electromagnetic spectrum. Therefore, far infrared waves are thermal. In 1989, NASA's COBE satellite detected *cosmic infrared background* (CIB) radiation in the near-infrared range. Some scientists believe that the CIB radiation represents a "core sample" of the universe. It contains the cumulative emissions of stars dating back to the epoch when these objects first began to form.

Ultraviolet waves – Ultraviolet waves are divided into three sub-ranges: the near ultraviolet (NUV) that lies close to visible light, the extreme ultraviolet (EUV) that lies close to X-rays, and the far ultraviolet (FUV) that lies between NUV and EUV. The ozone layer of the Earth's atmosphere absorbs high-energy radiation. Therefore, ultraviolet radiation can be detected only by using either rockets or satellites. Ultraviolet observations have contributed a great deal to the understanding of stellar evolution and formation, as well as the outer atmosphere of stars.

X-rays –X-rays were discovered by the German physicist Wilhelm Roentgen (1845-1923) in 1895, but it took another sixty-seven years before X-ray astronomy was born. The first rocket flight to successfully detect a cosmic source of X-ray emission was launched in 1962 by a group of American scientists headed by Riccardo Giacconi (b. 1931). Later it was found that the X-ray emission of this source is ten thousands times greater than its optical emission. It comes from sources which contain an extremely hot gas at temperatures from a million to one hundred million degrees Kelvin. The X-rays are emitted by compact celestial structures, such as neutron stars and black holes. Like the cosmic microwave background (CMB) radiation and the cosmic infrared background (CIB) radiation, the *cosmic X-ray background* (CXB) radiation has an extremely uniform distribution of intensity across the sky.

Gamma rays – Gamma rays are the most energetic form of electromagnetic waves; they are produced by the hottest regions of the Universe. They are also produced by such violent events as supernova explosions or the destruction of atoms, as well as during the decay of radioactive material. The other sources of gamma rays include neutron stars, pulsars, and black holes. In spite of the many discoveries of gamma-ray bursts, this enigmatic phenomenon continues to surprise astronomers. In December 1997, a gamma-ray burst was detected that turned out to be the largest explosion ever seen since the beginning of astronomic observations. Some astronomers think that this explosion was the result of a neutron star being drawn into a black hole.

Neutrinos - Contemporary theories define neutrinos as electrically neutral particles. Not affected by electromagnetic forces, they are able to pass through great expanses of matter without being affected by it. This makes the detection of neutrinos very difficult. In 1956 two American scientists, Clyde Cowan (1919-1974) and Frederick Reines (1918-1998), detected electron neutrinos near the nuclear plant on the Savannah River, in South Carolina. The first experiments to detect electron neutrinos produced by the Sun began in 1968. The experiments confirmed that the Sun produces electron neutrinos, but only about one-third of the number of neutrinos predicted by theory could be detected. This so-called "solar neutrino puzzle" was solved in 2001-2002 when the scientists working at the Sudbury Neutrino Observatory in Ontario, Canada, found strong evidence that the electron neutrino has the ability *to oscillate*, or transform into either muon or tau neutrinos.

Gravitational radiation – In 1915 Einstein predicted in his general theory of relativity that gravitational waves are emitted as a result of the fluctuation in the curvature of spacetime. He also predicted that gravitational waves propagate with the velocity of light. Because they are very weak, we can expect to directly observe only those gravitational waves on Earth that were generated by very distant and violent cosmological events, such as a collision of neutron stars or a collision of two super-massive black holes. Upon their arrival on Earth, the gravitational waves cause some changes in distances everywhere. But the magnitude of the changes is expected to be very small, on the order of about one thousandth of the diameter of a proton, making direct measurements of gravitational waves extremely difficult.

Although before 2006 gravitational radiation had not been directly detected, there is already significant indirect evidence for its existence. In 1974, two American astrophysicists, Russell Hulse (b. 1950) and Joseph Taylor, Jr. (b. 1941), discovered a binary pulsar made up of a pulsar and a black companion star. Although it was not understood at the time, this was the first of what are now called *recycled pulsars*: neutron stars that have been spun very fast around their centers by the accretion of mass onto their surfaces from a companion star. The orbit of this binary system is slowly shrinking as it loses energy through the emission of gravitational radiation. Hulse and Taylor convincingly demonstrated that the measurements of the shrinking rate of the binary system match the predictions from Einstein's theory with accuracy better than 1%.

1.10 Vortices in Nature

Are there any spiral entities in nature that we can observe without a use of either telescopes or microscopes? Of course, there are plenty of them. Among them are huge rotating storms periodically swirling over our oceans and lands, carrying with them the destruction of property and death. They are called *hurricanes* in the Atlantic and eastern Pacific Oceans, *typhoons* in the western Pacific and *cyclones* in the Indian Ocean.

Much smaller in size but more frequent are the *tornadoes* roaming over the lands. Many people had either direct or indirect encounter with these fearful forces of nature. Incredibly, this unbelievable force is created from weightless air by merely twisting it into a spiral shape, a puzzle still waiting for a complete answer.

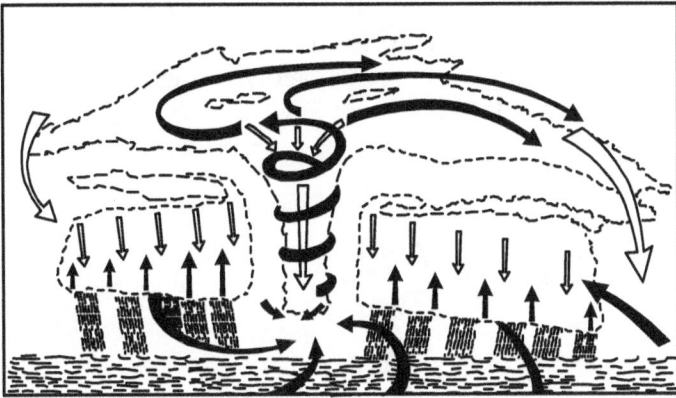

Figure 1.10. Formation of a hurricane by mixing warm air (solid arrows) with cold air (empty arrows). (Adapted from *Scientific American*, Oct. 2004).

The Norwegian meteorologist Vilhelm Bjerknes (1863-1951) was one of the first scientists who attacked this problem. In 1904 he became the first scientist to use successfully mathematical models for weather prediction. During the First World War he established a network of weather stations throughout Norway. Based on the information collected from these stations, a team that included Vilhelm Bjerknes, his son Jacob Bjerknes (1897-1975) and the Swedish meteorologist Tor Bergeron (1891-1977), developed a foundation of the first theory explaining a mechanism of precipitation of rain from clouds. These ideas were soon confirmed by the experimental studies and observations of the German meteorologist Walter Findeisen (1909-1945) and are now known as *Bergeron-Findeisen frontal theory*. This theory explains how hurricanes are generated over Atlantic Ocean, where warm and cold air masses meet (Fig. 1.10).

1.11 The Helical Theory of DNA

In their ground-breaking book *Understanding DNA – The Molecule and How It Works,* the British analytical engineer Christopher Calladine (b.1935) and the American-Australian molecular biologist Horace R. Drew (b.1955) clearly demonstrated that the double-helical structure provides the most favourable conditions for the DNA molecule to perform its important functions, while keeping the two chemically-opposite substances, the acids and the bases, extremely close to one another, but still separate (Fig. 1.11.1). Horace Drew kindly wrote for us below a brief history of the development of the helical theory of DNA between 1953 and 2013.

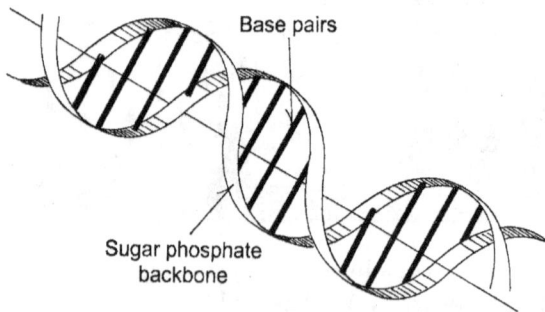

Figure 1.11.1. Schematic presentation of DNA.

DNA was soon confirmed to be the genetic material by biochemical methods. Certain triplets of bases on one strand of DNA could code for specific amino acids in a protein, say as "AAA" for lysine, "TTT" for phenylalanine, or "GCX" for glycine where X can be any of A, G, C or T. When cells divide, one strand of the double helix separates from the other, and then provides a template for the enzymatic synthesis of a new complementary strand. Simple mutations in the double helix (say from A to G, C or T), at any of six billion base pairs in a human genome, were envisaged to facilitate Darwinian evolution by random changes followed by natural selection.

Several different helical forms of DNA were subsequently seen by the x-ray diffraction of pulled fibers. Between 1952 and 1980, these helical forms were identified as "A, B, C, D or E" by the British biophysicist Rosalind Franklin (1920-1958), the American biochemist and biophysicist Robert Langridge (b.1933) and the Scottish molecular biologist Struther Arnott (1934-2013). How might such alternative forms of DNA be related to the model made by the American microbiologist James Watson (b. 1928) and the English molecular biologist Francis Crick (1916-2004)? In 1980, after making single crystals from short pieces of DNA which had been synthesized chemically, two groups at MIT or Caltech, led by American x-ray crystallographers Alexander Rich (b.1924) or Richard Dickerson (b.1931) respectively, found a left-handed form of DNA which they called "Z". Other studies of right-handed DNA showed that certain sequences of bases A, G, C, or T could influence the helical structure to some extent, but no one understood why.

Also no one understood how DNA interacts specifically with certain proteins called "transcription factors" or "repressors" to turn genes "on or off", or how DNA wraps tightly about certain proteins called "histones" in the cell nucleus. Still other studies by the American microbiologist Donald Crothers (b.1937) and others suggested that long pieces of DNA could be naturally curved due to their base sequence in water solution without protein, and again this was poorly understood. Amidst such a proliferation of new data, Christopher Calladine thought in 1983 that it might be useful to describe the DNA double helix in terms of "six degrees of freedom" which had originally been studied by Leonard Euler, namely three of translation and three of rotation (Fig. 1.11.2).

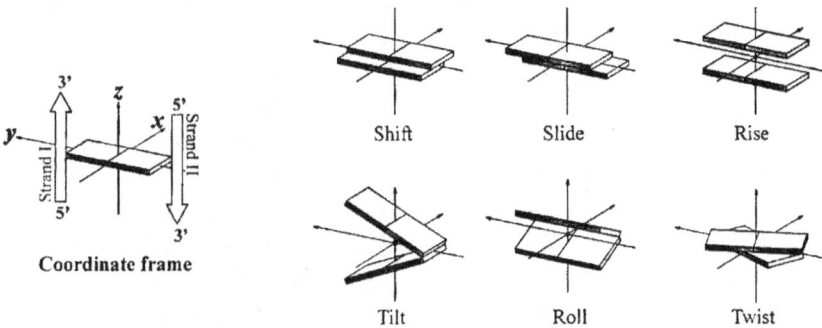

Figure 1.11.2. The DNA double helix in terms of "six degrees of freedom" (three of translation and three of rotation).

As applied to DNA, those six parameters are called "shift, slide, rise" for translation, or "tilt, roll, twist" for rotation. A careful study of atomic coordinates, taken from the x-ray diffraction of single crystals of DNA, next showed that the "B-to-A" transition of DNA seen by Rosalind Franklin in the 1950's was a combination of just two Euler parameters "roll" and "slide"(Fig. 1.11.3).

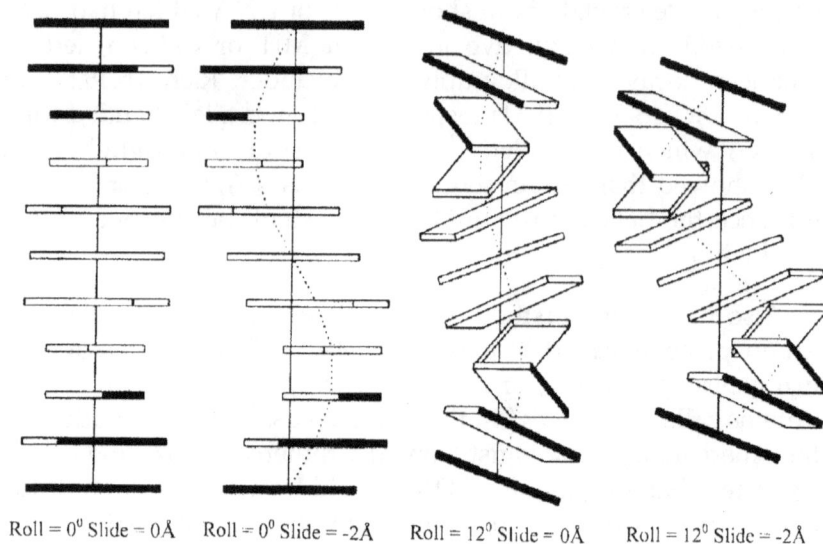

Roll = 0° Slide = 0Å   Roll = 0° Slide = -2Å   Roll = 12° Slide = 0Å   Roll = 12° Slide = -2Å

Figure 1.11.3. The "B-to-A" transition of DNA as a combination of just two Euler parameters "roll" and "slide."

Christopher Calladine had previously studied the dynamics of bacterial flagella, which may adopt a variety of helical forms, and where such concerns are equally applicable. When complexes of DNA with various proteins were studied in the 1980's or 1990's, by the x-ray diffraction of single crystals, often we saw that the double helix would curve tightly about a protein. Atomic coordinates became available for many examples. On careful study of those coordinates, many workers found that the DNA double helix would curve smoothly about a protein, using a simple "roll" motion which varied as a Fourier wave (due to the helix twist), as shown in case (Fig. 1.10.4d) below on the right.

The bases A, G, T or C are not very soluble in water, so smooth bending by "roll" (without any "kinks") provides the lowest-energy solution of how to curve DNA without exposing its hydrophobic surfaces to water. In mathematical language, we can say that the local curvature of DNA may be approximated by a Fourier transform of its "roll" angles. This is similar to what Erwin Schroedinger postulated in 1944, when he proposed in his book *"What is Life?"* that the genetic material (not known then) would be an "aperiodic crystal".

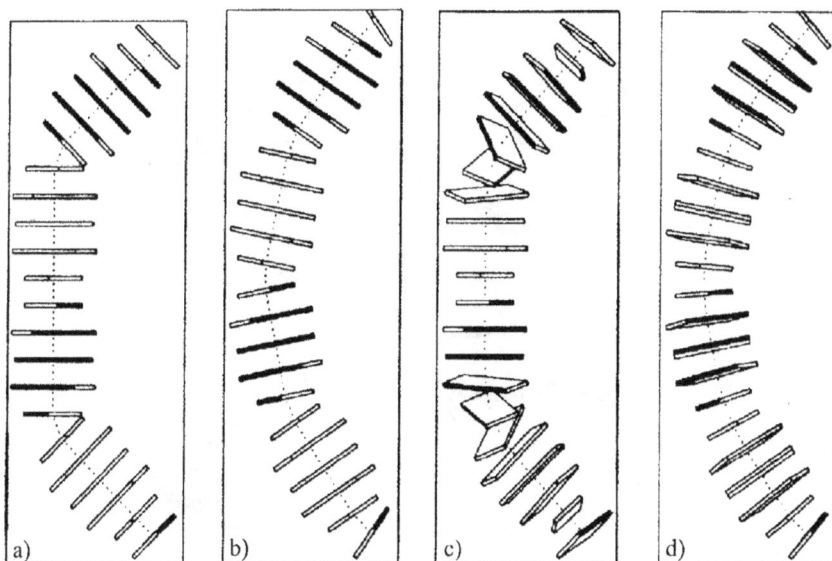

Figure 1.11.4. The DNA double helix curves smoothly about a protein, using a simple "roll" motion (d).

Such "smooth rolling motion" is especially true for a complex between histone proteins in the cell nucleus, and 145 base pairs of DNA wrapped tightly around it into almost two super-helical turns, in what is called the "nucleosome core particle" of chromosomes. The atomic coordinates of such a complex were determined by the British biophysicist Aaron Klug (b.1926), Italian biochemist Daniela Rhodes (b. 1950) and American x-ray crystallographer Tim Richmond (b. 1953) in a series of investigations conducted between 1978 and 1990.

This "roll-slide-twist" model for DNA is just a geometrical abstraction, because it does not address any underlying chemical features of the bases A, G, C or T, or of their attached sugars and phosphates, which might give rise to observed behaviours. That deeper level was addressed by a New Zealand-British chemist Chris Hunter (b.1965) in the 1990's, when he made new theoretical calculations which included partial electrical charges on the base pairs (Fig. 1.11.5). We can see that there are negative partial charges (-) in certain places, or positive partial charges (+) in other places. By including detailed chemical information in his models, Hunter was able to successfully model the "roll-slide-twist" behaviour on an atomic or chemical scale.

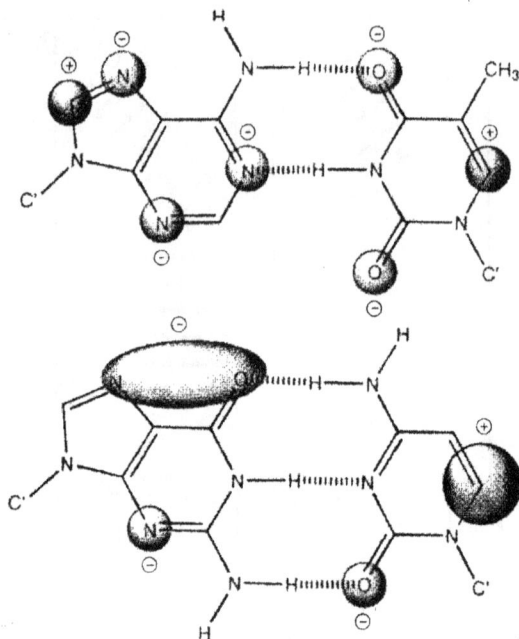

Figure 1.11.5. Partial electrical charges on the base pairs.

Outside of his house on Portugal Place in Cambridge, Francis Crick erected a large "wire helix", because he became convinced that helices would be a fundamental aspect of all sciences on Earth. His reasoning was that, if you take any small building block, and then join a series of such blocks end-to-end, you will always form in three dimensions some kind of "helix", by the combination of translation and rotation. In higher dimensions we might see four or five-dimensional "helices" of even more complex shapes, although we do not understand that subject very well at present.

Currently particle physicists are studying "points" or "strings" in an attempt to explain subatomic particles in as many as 11 dimensions. It might not be a bad idea to study "helices" while they are so engaged.

CHAPTER 2

MULTI-DIMENSIONAL SPACETIME

The dimensions of space are most frequently defined as the minimum number of coordinates needed to specify each point within it.

2.1 Euclid's Multi-Dimensional Spaces – Here is how to create multi-dimensional spaces in accordance with Euclid's geometry.

The point m (Fig.2.1) has no dimensions at all. Move the point m a distance x along a straight path (a) and you will obtain a line that will have one dimension, length x. Moving this line a distance y in the direction perpendicular to the line will produce a two-dimensional plane (b) with the sides x and y. Now move this plane a distance z in the direction perpendicular to the plane and you will produce a three-dimensional solid (c) with the sides x, y and z.

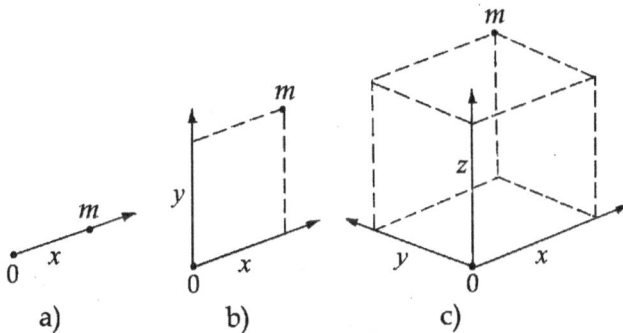

Figure 2.1. Spaces of one (a), two (b) and three (c) dimensions.

Many ancient Greek scientists, including Aristotle and Ptolemy, believed that the fourth dimension is impossible.

2.2 Fitch's Four-Dimensional Hypercube – The four-dimensional *hypercube* was proposed by a member of the Corps of Engineers, U.S.A. Lieutenant-Colonel Graham Denby Fitch. He described his idea in the *Scientific American* July issue of 1909.

Fitch produced the hypercube (Fig. 2.2) by moving each corner of a 3-dimensional cube *ABCDEFGH* in parallel directions *Aa, Bb, Cc, Dd, Ee, Ff, Gg* and *Hh.* These directions are supposed to be perpendicular to our 3-dimensional space. The four-dimensional hypercube was complicated and very difficult to visualize. One may rightfully wonder how complicated must be the spaces of higher dimensions, and could be they "visualizable" at all.

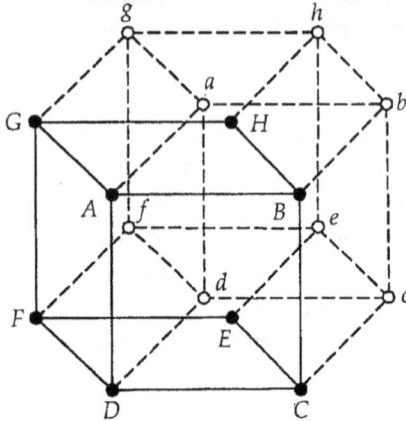

Figure 2.2. The 4-dimensional hypercube.
Adapted from H.P. Manning (1960).

2.3 Riemann's Multi-dimensional Space - The German mathematician Georg Riemann (1826-1866) defined the dimensions of a space by a degree of the space curvature. The higher degree of curvature of space the greater number of coordinates will be required to locate a point in this space.

For instance, the position of a point *A* on a two-dimensional curved line (Fig. 2.3) is determined by two coordinates, *x* and *y,* while three coordinates *x, y,* and *z* are needed to define a position of a point *A* on a three-dimensional sphere. In both cases the modulus of the vectors defining the location of the point *A* is given by similar equations:

$$z^2 = x^2 + y^2$$
$$u^2 = x^2 + y^2 + z^2$$

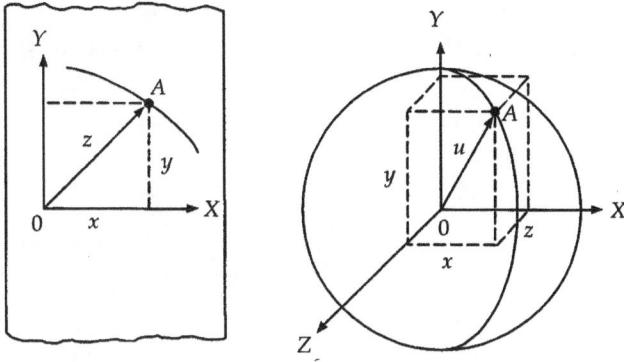

Figure 2.3. Riemann curved spaces of two and three dimensions.

A curved space of any higher order can be described by adding extra terms into the last equation.

2.4 Minkowski's Spacetime - The Russian-German mathematician Hermann Minkowski (1864-1909) made his outstanding contribution to relativity by uniting space and time.

When time and space are considered independent of each other, the location of an object in a three-dimensional space is defined by three dimensions x, y, and z. To connect space with time t, Minkowski added one more dimension ict in which $i = \sqrt{-1}$ is the imaginary number. Thus, the modulus of the vector u defining the location of a point on a curved space was defined by the modified Riemann equation:

$$z^2 = x^2 + y^2 - (ct)^2$$
$$u^2 = x^2 + y^2 + z^2 - (ct)^2$$

2.5 Einstein's Spacetime - The German-Swiss-US theoretical physicist Albert Einstein (1879-1955) expanded the application of the term spacetime in his General Theory of Relativity by using a special branch of mathematics called *tensor calculus*.

The result was a groundbreaking departure from the Newtonian interpretation of gravity. Gravitation was no longer a force but an intrinsic curvature of spacetime. It was the curved space-time around the Sun that attracted the planets. In curved space-time the motions of the planets were represented by the shortest distances, called *geodesics*. Figure 2.5.1 illustrates the curvature of spacetime according to Einstein.

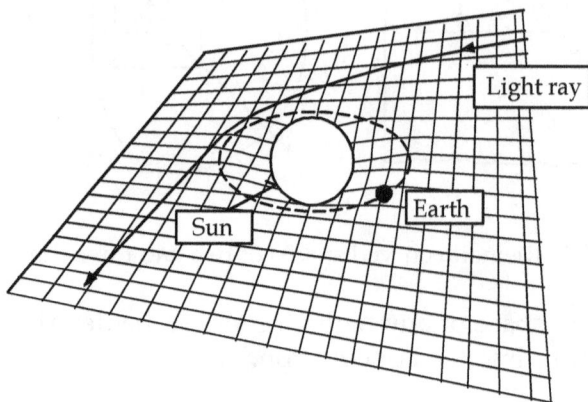

Figure 2.5.1. Einstein's curved spacetime.

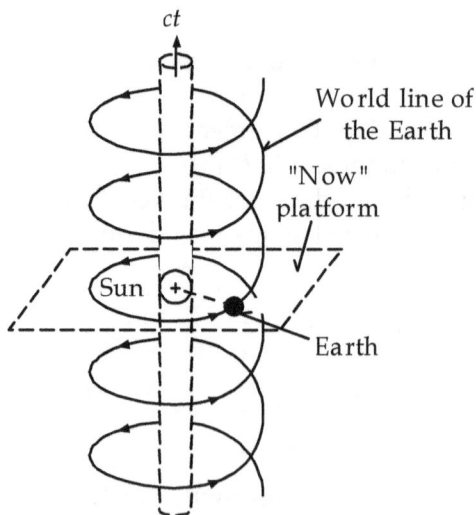

Figure 2.5.2.
The space-time of
the Sun and the Earth.
Adapted from
B. Hoffmann (1972).

It is difficult to illustrate the curvature of time diagrammatically. In Figure 2.5.2, we omitted the curvature of time to simplify the illustration of Einstein's space-time. The double-dotted line represents the sun's world line as it moves upwards on the page. The world line of a planet is represented by a helical spiral that is a geodesic in the curved space-time associated with the Sun. Imagine that the Earth is located at this moment on a platform indicated as the "now" platform. As time passes, the platform will move upwards on this page a distance *ct*.

2.6 Kaluza's Multi-Dimensional Space - The German physicist Theodor Kaluza (1885-1954) presented the multi-dimensional space by wrapping up a space of a higher dimension around a space of a lower adjacent dimension.

Consider that a few hundred feet of garden hose is stretched across a canyon, and you view it from a far distance away. From this distance the hose will appear as one-dimensional string. Therefore, the ant located on the hose would have only one dimension in which to walk: the left-right dimension along the hose's length (Fig. 2.6).

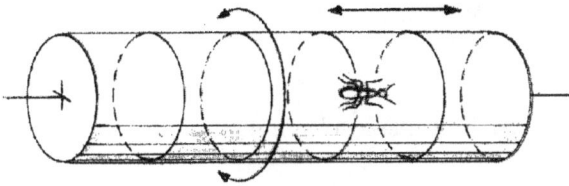

Figure 2.6. Kaluza's multi-dimensional space.

From a magnified perspective, we will see that besides the left-right dimension along the hose's length there is the second dimension along which the ant can walk, either clockwise or counterclockwise. We may say that the second dimension of the hose as "wrapped up" around the one-dimensional material. Similarly, it is possible to describe mathematically a three-dimensional hose "wrapped up" around the two-dimensional material, and so on and so on.

2.7 Spiral Principle – The spiral principle is a purely abstract proposition of the SST used to discover a multi-dimensional spacetime called *helicola*. This principle simply states:

Every line is a three-dimensional spiral.

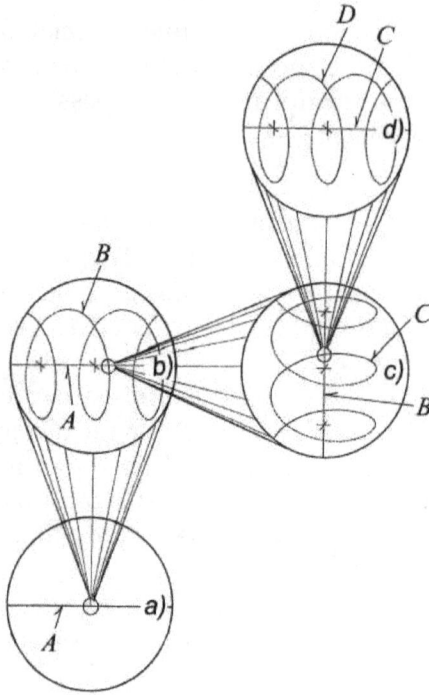

Figure 2.7.1. Multiple-level spiral of diminishing dimensions.

By following this principle, one will discover a multiple-level spiral of diminishing dimensions (Fig. 2.7.1). Start from a straight line A (Fig. 2.7.1a). Then apply the spiral principle to discover that the line A, upon a closer examination, looks like a spiral B wound around the line A (Fig. 2.7.1b). But, the windings of the spiral B are also made of a line. Therefore, according to the spiral principle, this line, under a closer examination, appears as a spiral C wound around the spiral B (Fig. 2.7.1c). A further more detail examinations will reveal a spiral D wound around the spiral C (Fig. 2.7.1d), and so on.

The Spiral Principle allows us also to discover a multiple-level spiral of enlarging dimensions (Fig. 2.7.2). Start from a straight line A (Fig. 2.7.2a). According to the spiral principle, when looking at the line A from a long distance away, it will appear as a part of a large spiral A wound around a line B (Fig. 2.7.2b). Similarly, when looking from a farther distance away from the line B, it will appear as a spiral B wound around a line C (Fig. 2.7.2c). After moving further away from the line C, it will reveal itself as a spiral D wound around a line D (Fig. 2.7.2d), and so on.

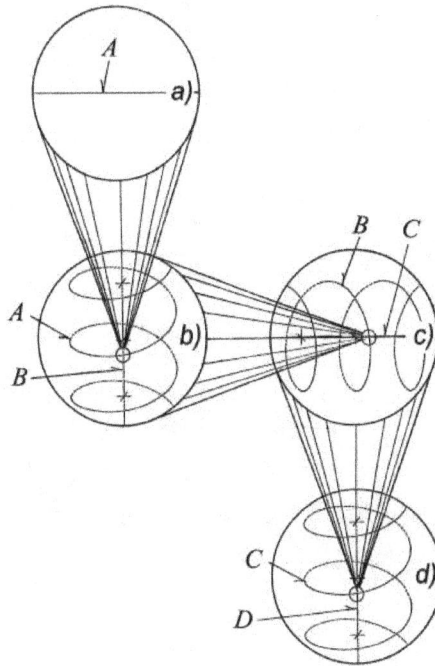

Figure 2.7.2. Multiple-level spiral of enlarging dimensions.

2.8 Helicola is the universal multi-dimensional spacetime which geometry is based on the spiral principle.

At each level of helicola there are two strings, a *leading string* and a *trailing string*. In Figure 2.7.1b, the leading string A is accompanied by a trailing string B. At the next level of helicola shown in Figure 2.7.1c, the trailing string B of the previous level of helicola becomes a leading string, and it is accompanied by the trailing string C. Similarly, at the next level of helicola shown in Figure 2.7.1d, the trailing string C of the previous level of helicola becomes a leading string, and it is accompanied by the trailing string D, and so on. The dimensions of helicola are defined by the number of parameters required to locate a point on its outer string.

2.9 Helicola of Odd Dimensions

Figure 2.9 shows the examples of helicola of one, three and five dimensions.

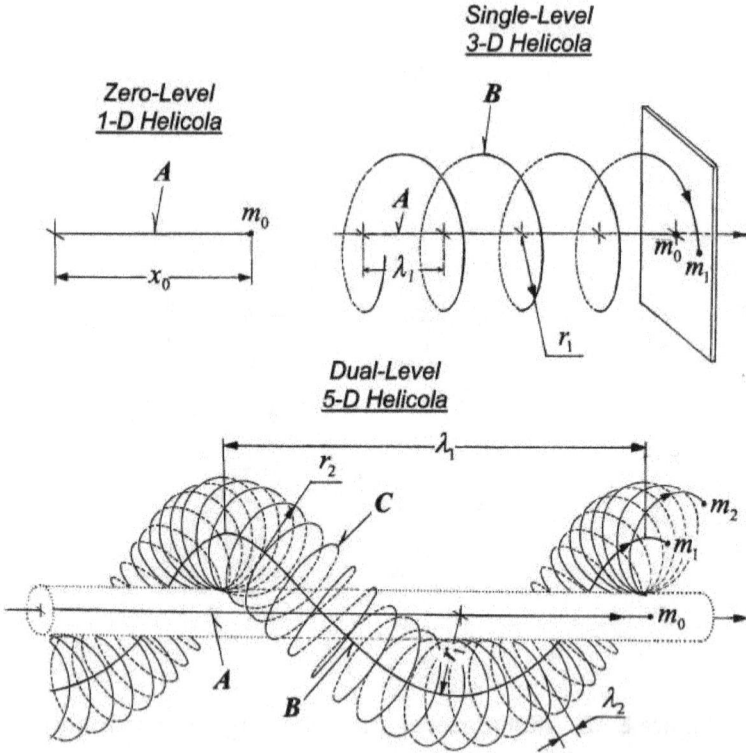

Figure 2.9. The odd-dimensional helicola.

- The 1-D zero-level helicola contains only a straight-line leading string A, and may be viewed as an incomplete helicola. Only one parameter x_0 is required to define the location of the point m_0 on this string.

- In the 3-D single-level helicola, the point m_1 belongs to the trailing string B, and it propagates synchronously with the point m_0. To define the location of the point m_1 it is necessary to know two additional parameters of the trailing string B, the wavelength λ_1 and the radius r_1.

- In the 5-D dual-level helicola, the point m_2 belongs to the second trailing string C, and it propagates synchronously with two points, m_1 and m_0. To define the location of the point m_2 it is necessary to know two more parameters of the second trailing string C, the wavelength λ_2 and the radius r_2.

2.10 Helicola of Even Dimensions – Figure 2.10 shows the examples of helicola of two, four and six dimensions.

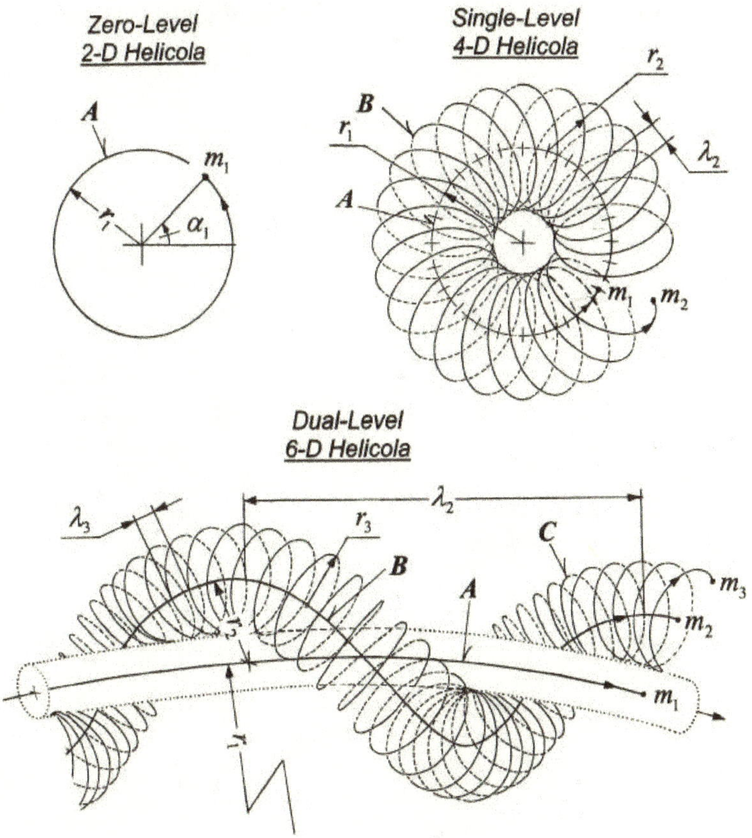

Figure 2.10. The even-dimensional helicola.

- In the 2-D zero-level helicola, the point m_1 belongs to the circular leading string A with the radius r_1. To define the location of the point m_1 it is necessary to know two parameters of leading string A, the radius r_1 and the angle α_1.

- In the 4-D single-level helicola, the point m_2 belongs to the toroidal trailing string B, and it propagates synchronously with the point m_1. To define the location of the point m_2 it is necessary to know two additional parameters of the trailing string B, the wavelength λ_2 and the radius r_2.

- In the 6-D dual-level helicola, the point m_3 belongs to the second trailing string C, and it propagates synchronously with two points, m_1 and m_2. To define the location of the point m_3 it is necessary to know two more parameters of the second trailing string C, the wavelength λ_3 and the radius r_3.

CHAPTER 3

STATIC & DYNAMIC HELICAL SPIRALS

We do not have to go too far to see how helical spiral looks like. Many of us saw it already in a popular toy called "slinky." We will construct two kinds of helical spirals, static and dynamic. Nothing moves in the static helical spiral, and we can easily make it. The dynamic spiral is formed by the motion of its strings. Therefore, we have to use our imagination to envision it.

3.1 Static Single-Helical Spiral - Here is one of the ways of making a static single-helical spiral. It involves using three easy steps.

Step 1 – Take a round object like a beverage can shown in Figure 3.1a. In geometry, this round object is known as a *cylinder*. They describe the cylinder geometry by using its two dimensions: the radius r_1 and the height λ_1.

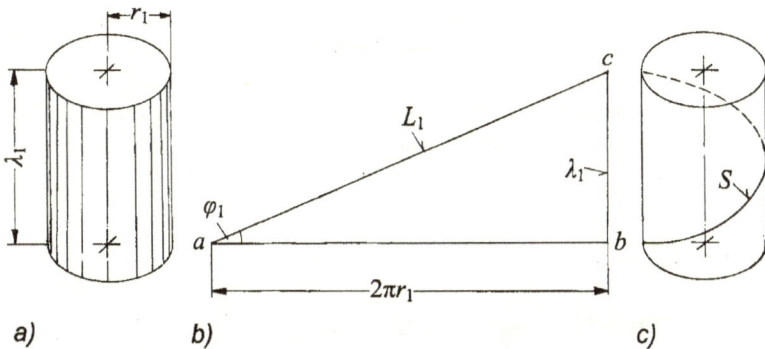

Figure 3.1. Construction of a static single-helical spiral.

Step 2 - Cut a right triangle *abc* out of a piece of paper as shown in Figure 3.1b. Make the length of one side *ab* equal to the perimeter of the can $2\pi r_1$ and the length of the other side *bc* equal to the height of the can λ_1

Step 3 - Finally, wrap the prepared triangular piece of paper around the can as shown in Figure 3.1c. As the side *ab* will wrap around the bottom of the can, the diagonal *ac* will appear on the can surface as one winding of a helical spiral.

We will call the cylinder radius r_1 the *radius of helical spiral* and the can height λ_1 the *wavelength of helical spiral*. These are the basic spiral parameters. There are three derivative parameters of the helical spiral, the *length of one spiral winding* L_1, the *spiral steepness angle* φ_1 and the *spiral volume* U_1. These three parameters are called *derivative*, because they can be derived from the basic parameters. Main space parameters of the static single-helical spiral are:

L_1 = spiral length
r_1 = spiral radius
U_1 = volume
λ_1 = wavelength
φ_1 = steepness angle.

Based on Figure 3.1b, we can determine the spiral length L_1 by using the Pythagorean Theorem:

$$L_1 = \sqrt{(2\pi r_1)^2 + \lambda_1^2} \qquad (3.1\text{-}1)$$

We can determine the cosine of the spiral steepness angle φ_1 by using conventional trigonometry,

$$\cos \varphi_1 = \frac{2\pi r_1}{L_1} \qquad (3.1\text{-}2)$$

The volume U_1 is equal to the volume occupied by the static single-helical spiral within one wavelength λ_1, it is equal to:

$$U_1 = \pi r_1^2 \lambda_1 \qquad (3.1\text{-}3)$$

Importantly, because the leading string of the helical spiral is in the form of a straight line, the values of spacetime parameters of the helical spiral are independent on the position of a point of measurement located along the trailing string.

3.2 Dynamic Single-Helical Spiral - To construct the dynamic single-helical spiral depicted in Figure 3.2, you may

simply use tips of pointing fingers of your hands. Call the pointing finger tip of one of your hands the point m_0 and the pointing finger tip of the other hand the point m_1. Then use your imagination and follow the three easy steps shown below.

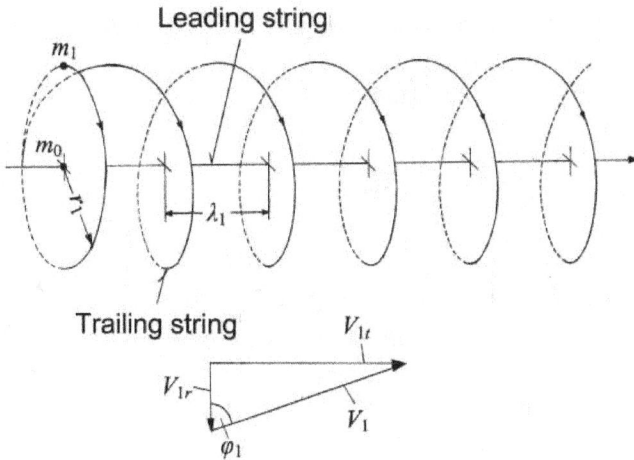

Figure 3.2. Construction of a dynamic single-helical spiral.

Step 1 – Stand up and stay firmly on the floor. Use a pointing finger of one of your hands to locate the point m_0 in front of you. Use a pointing finger of your other hand to locate the point m_1 at the distance r_1 from the point m_0.

Step 2 – Move the point m_1 around the stationary point m_0 along a circle with the radius r_1. Imagine a circular trace left in the air by the moving point m_1. Make this circular trace with any speed you feel comfortable with, keeping in mind that this speed is called the *rotational velocity* V_{1r}.

Step 3 – As you continue making the circular trace left by moving point m_1, begin walking forward by maintaining the same distance from the point m_0 in front of you. Move forward with any speed you feel comfortable with, keeping in mind that this speed is called the *translational velocity* V_{1t}.

An imaginary straight-line trace left by the point m_0 is called the *leading string* of the single-helical spiral. As the point m_1 rotates around the point m_0 and simultaneously moves forward together with the point m_0, it leaves an imaginary trace called the *trailing string* of single-helical spiral. The trailing string of this spiral will propagate along its spiral path with the *spiral velocity* V_1. We can describe the relationship between the spiral velocity V_1 and the velocities of its two components, the rotational velocity V_{1r} and the translational velocity V_{1t}, by using the Pythagorean Theorem.

Notably, the spiral velocities are the *spacetime parameters* expressed in the units of space and time, such as meters per second (*m/s*). The other two parameters of the dynamic single-helical spiral, the spiral period T_1 and the spiral frequency f_1, are the *time parameters* expressed in the units of time, such as second (*s*) or 1/second (*s⁻¹*).

The space parameters of the dynamic single-helical spiral are the same as these of the space parameters of the static single-helical spiral and its space equations are defined by Eqs. (3.1-1) – (3.1-3). The spacetime parameters of this spiral are:

f_1 = frequency
T_1 = period
V_1 = spiral velocity
V_{1r} = rotational velocity
V_{1t} = translational velocity.

From Figure 3.2 and based on Pythagorean Theorem, the spiral velocity V_1 is related tor the rotational velocity V_{1r} and the translational velocity V_{1t} by the equation:

$$V_1 = \sqrt{V_{1r}^2 + V_{1t}^2} \qquad (3.2\text{-}1)$$

The period T_1 is the time required for the dynamic single-level helical spiral to propagate through one its full cycle. Based on this definition, the period T_1 is equal to:

$$T_1 = \frac{L_1}{V_1} = \frac{2\pi_1}{V_{1r}} = \frac{\lambda_1}{V_{1t}} \qquad (3.2\text{-}2)$$

The frequency f_1 is the inverse of the period T_1. Therefore, we obtain from Eq. (3.2-2) :

$$f_1 = \frac{V_1}{L_1} = \frac{V_{1r}}{2\pi_1} = \frac{V_{1t}}{\lambda_1} \qquad (3.2\text{-}3)$$

Importantly, because the leading string of the static single-helical spiral is in the form of a straight line, the values of space-time parameters of this spiral are independent on the position of a point of measurement located along the trailing string.

3.3 Dynamic Double-Helical Spiral - The dynamic double-helical spiral is made up of two dynamic single-helical spirals. It is wide-spread in the universe.

Consider, for instance, a binary star that is a star system consisting of two stars orbiting around their common center of gravity. The paths of these two stars correspond to the dynamic helical spiral. To construct the dynamic double-helical spiral depicted in Figure 3.3, you may use a pencil. Call one end of the pencil the point *m* and the other end the point *n*. Then use your imagination and follow the three easy steps shown below.

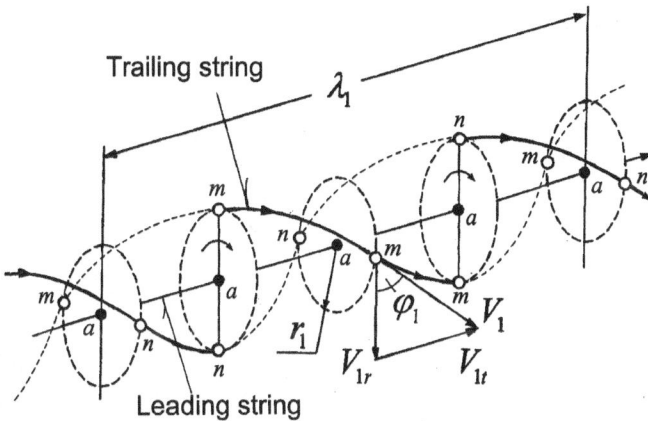

Figure 3.3. Construction of a dynamic double-helical spiral.

Step 1 - Stand up and stay firmly on the floor. Use fingers of one of your hand to hold the pencil at its middle point *a*. Keep the pencil in front of you.

Step 2 - Move the pencil around its middle point *a*. Imagine a double-circular trace left in the air by the pencil's end points *m* and *n*. Make this double-circular trace with any speed you feel comfortable with, just keep in mind that this speed is called the *rotational velocity* V_{1r}.

Step 3 - As you continue making the double-circular trace left by the pencil's end points m and n, begin walking forward by maintaining the same distance from the pencil in front of you. Move forward with any speed you feel comfortable with, keeping in mind that this speed is called the *translational velocity* V_{1t}.

An imaginary straight-line trace left in the air by the pencil's end point a is called the *leading string* of the double-helical spiral. The double-helical trace left in the air by the pencil's end points m and n are called the *trailing string* of this spiral. The trailing string will propagate along its spiral path with the *spiral velocity* V_1.

Notably, the spacetime properties of each branch of the dynamic double-helical spiral are described by the same equations that are applied to the dynamic single-helical spiral. Space-wise the branches are separated by a distance of one half of the wavelength λ_1. Time-wise the branches are separated by one half of the period T_1.

CHAPTER 4

UNIVERSAL HELICOLA

As many other theories, The UST had a clear starting point. It all started in December 1992 during a dinner conversation with my son Gene, who was then finishing his last year of high school. While looking at the crumb of bread, Gene made a casual observation that there is probably another world inside that piece. Even though I heard this notion more than once before, somehow it registered in my mind differently this time. I started to wonder if there is any geometrical shape that would perfectly represent the infinitude of the world and its repetitious character in both enlarging and diminishing directions. This shape turned out to be the endless multidimensional helical spiral which I called the *helicola.*

At first I thought that the helicola was a purely abstract spacetime construction having nothing to do with our universe. But, a few days later I realized that the helicola outlines the paths of real celestial bodies. The first example that came to my mind was a helicola created by the paths of the center of our galaxy the Milky Way, the Sun, the Earth and the Moon.

4.1 Spiral Path of the Sun – As shown in Figure 4.1, the helicola of the first level describes paths of the Sun and the center of the Milky Way. The leading string 1 of helicola represents a path of the center of the Milky Way toward the constellation Hydra at the translational velocity V_{1t}. The Sun is located at the distance r_1 from the galaxy center. As the Milky Way moves toward the constellation Hydra, the Sun orbits the galaxy center at the rotational velocity V_{1r}.

A combination of the rotational motion of the Sun with the translational motion of the galaxy center makes the Sun moving along a spiral path of the trailing string 1 of helicola. The Sun spiral velocity V_1 is equal to the geometrical sum of the rotational velocty V_{1r} and translational velocity V_{1t}. The trailing string 1 of the first level of helicola also serves as a leading string 2 of the second level of helicola described in the next section.

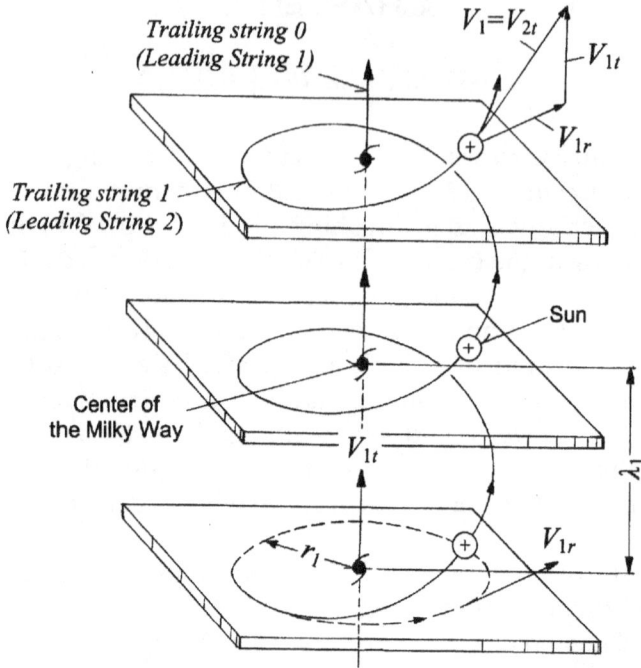

Figure 4.1. Spiral path of the Sun around the center of Milky Way.

4.2 Spiral Path of the Earth - The helicola of the second level
(Fig. 4.2) describes paths of the Earth and the Sun. At this
level, the leading string *2* is a former trailing string *1* of the
first level of helicola; it outlines the path of the Sun along its
spiral path at the translational velocity V_{2t} that is equal to the
spiral velocity V_1 of the trailing string *3*.

A combination of the rotational motion of the Earth and the
translational motion of the Sun makes the Earth moving along a
spiral path of the trailing string *2* of helicola. The Earth spiral velocity V_2 is equal to the geometrical sum of the rotational velocity
V_{2r} and translational velocity V_{2t}. The trailing string *2* of the second level of helicola also serves as a leading string *3* of the third
level of helicola described in the next section, so $V_{2t} = V_1$.

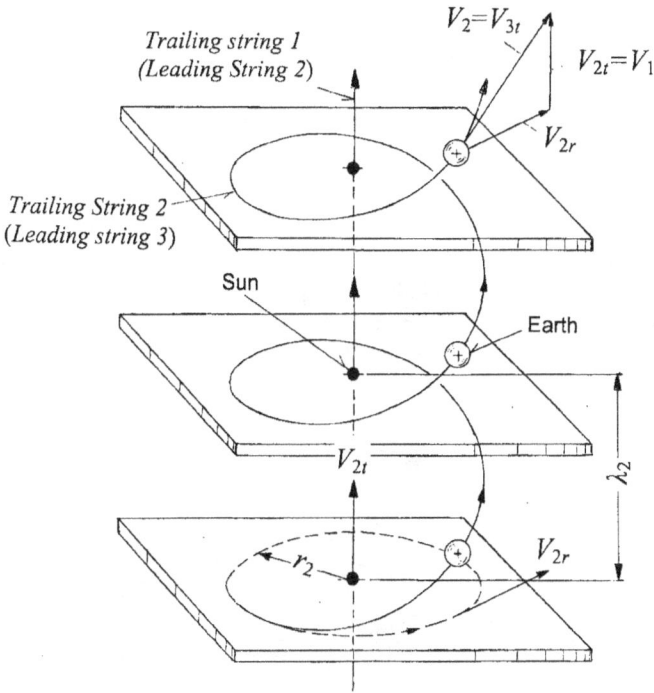

Figure 4.2. Spiral path of the Earth around the Sun.

4.3 Spiral Path of the Moon - The helicola of the third level (Fig. 4.3) describes the paths of the Moon and the Earth. At this level, the leading string *3* is a former trailing string *2* of the second level of helicola; it outlines a path of the Earth along its spiral path at the translational velocity V_{3t} that is equal to the spiral velocity V_2 of the trailing string *3*.

A combination of the rotational motion of the Moon and the translational motion of the Earth makes the Moon moving along a spiral path of the trailing string *3* of helicola. The Moon spiral velocity V_3 is equal to the geometrical sum of the rotational velocity V_{3r} and translational velocity V_{3t}. The trailing string *3* of the third level of helicola also serves as a leading string *4* of the fourth level of helicola that could be a path of a satellite orbiting the Moon.

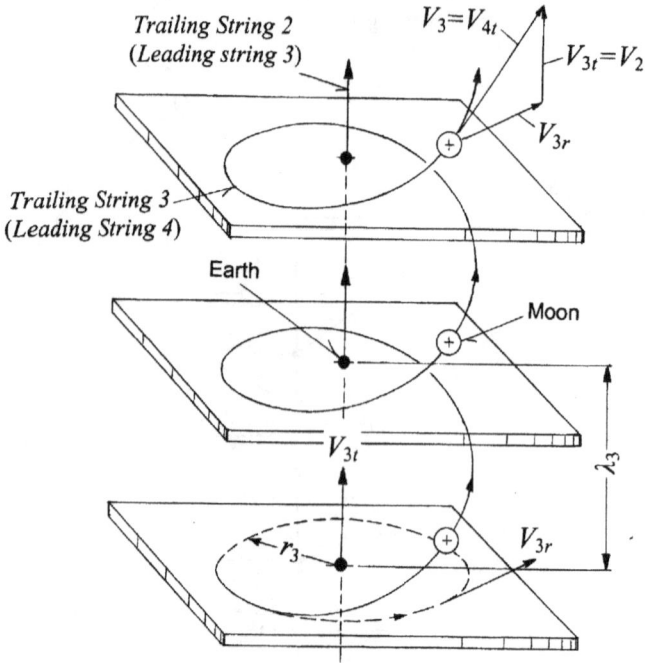

Figure 4.3. Spiral path of the Moon around the Earth.

The traces left by the center of the Milky Way, the Sun, the Earth and the Moon form a three-level helicola. I thought that the levels of the helicola will increase as we will consider the motion of the centre of the Milky Way around the center of even more gigantic entity, and then the motion of the center of that gigantic entity around the center of the next super-gigantic entity, and so on and so on.

4.4 Solar Planetary System - After the discovery of the celestial helicola it became obvious to me that the commonly-known illustration of the solar system in which the stationary Sun is surrounded by the planets moving along elliptical paths (see Figure 4.4a) is outdated.

Proposed by the German astronomer and mathematician Johannes Kepler more than five centuries ago, it was based on the assumption that the stationary Sun was located at the center of the universe. Figure 4.4b provides us with a contemporary view of our planetary system in which the planets spiral around the Sun while the Sun orbits the center of the Milky Way.

Ironically, this spiral view of our planetary system was known to the astronomers and astrophysicists since they discovered many years ago that the Sun moves around the center of the Milky Way. However, the image of elliptical planetary orbits still remains in the minds of many ordinary people, and many publications still show outdated illustrations of our solar system.

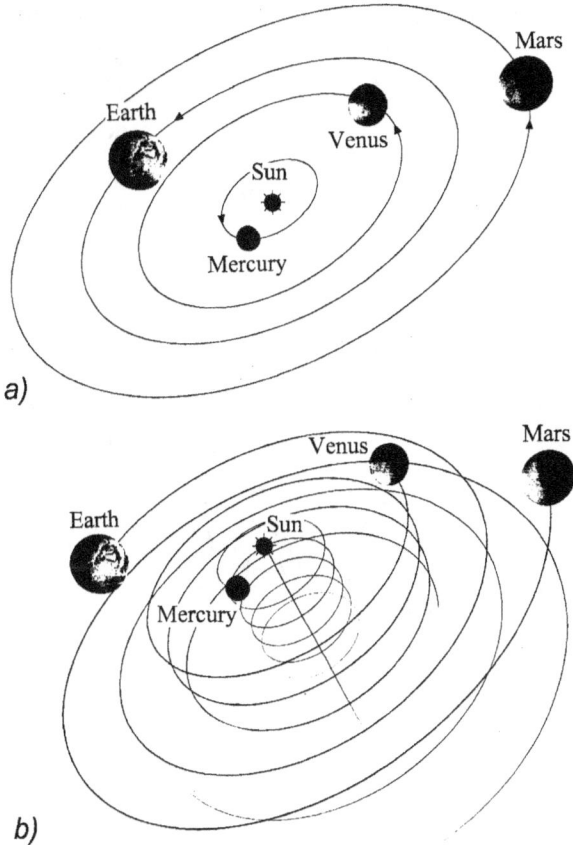

Figure 4.4. Presentations of a solar planetary systems:
a) outdated and b) updated.

4.5 Helicola of Macro- & Micro-Worlds - The discovery of the celestial helicola made me think that the spiral motion may be at the core of creation and existence of the entire universe.

I began to view hurricanes, tornadoes, whirlpools and all other spiral structures, including DNA, as merely transitional entities that may exist in nature all the way down to the prime elements of nature forming elementary particles. I wondered if the spacetimes of the prime elements of nature are merely particular cases of helicola. To illustrate this proposition I made a picture of a well-known structure of a hydrogen atom in which the atomic electron B orbited the proton A. In this picture shown in Figure 4.5, the propagation of the atomic electron was accompanied by a toroidal spiral. This spacetime arrangement was a particular case of helicola, the single-level 4-D helicola shown in Figure 2.10. In this helicola, the path of the electron was leading string and the path of the toroidal spiral the trailing string.

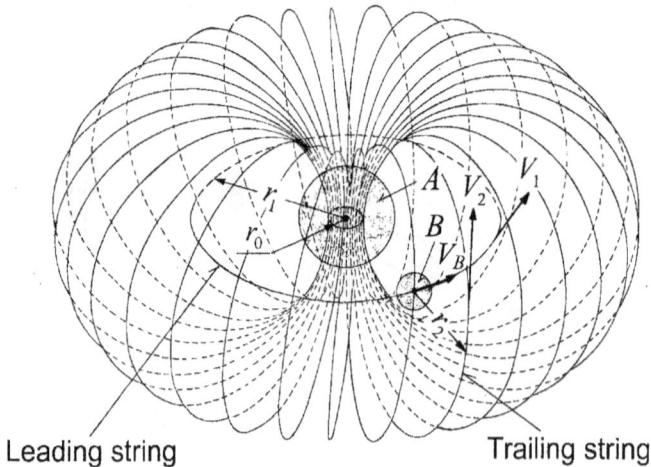

Leading string   Trailing string

Figure 4.5. Toroidal spiral accompanying atomic electron in the hydrogen atom.

The helicola turned out to be very suitable for modeling the spacetime properties of elementary particles. The single-level 4-D helicola provided required spacetime properties for the constituents of *elementary matter particles* called the *toryces*. Another particular case of helicola, the dual-level 5-D helicola shown in Figure 2.9 provided required spacetime properties of constituents of *elementary radiation particles* called the *helyces*.

CHAPTER 5

BASIC CONCEPT OF TORYX

The name "toryx" reflects the geometry of its spiral shape that appears like a conventional torus. You do not have to be a great scientist to envision the toryx. Simply use a very common toy known to many of us as "slinky" with approximately thirty windings. Bend the cut part into a shape of a doughnut and clip its end windings together as shown in Figure 5.

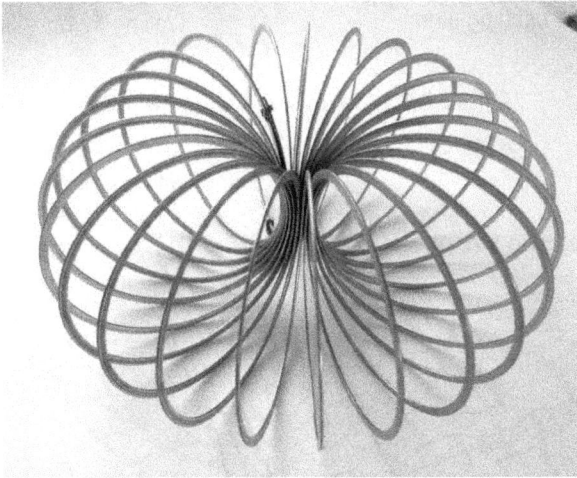

Figure 5. Mechanical model of toryx.

5.1 Toryx Basic Structure - Toryx consists of three main parts: a double-circular leading string, a double-toroidal trailing string and a spherical membrane as shown in Figs. 5.1.1 – 5.1.4. Notice that for the sake of simplicity these figures show only one branch of leading and trailing strings.

Double-circular leading string – The double-circular leading string of a toryx is a particular case of the dynamic double-helical spiral shown in Figure 3.3 in which the translational velocity V_{1t} is equal to zero and the rotational velocity V_{1r} is equal to the spiral velocity V_1. Consequently, the double-circular leading string appears like two circular traces left by the moving points m and n shown in Figure 5.1.1.

Basic Concept of Toryx

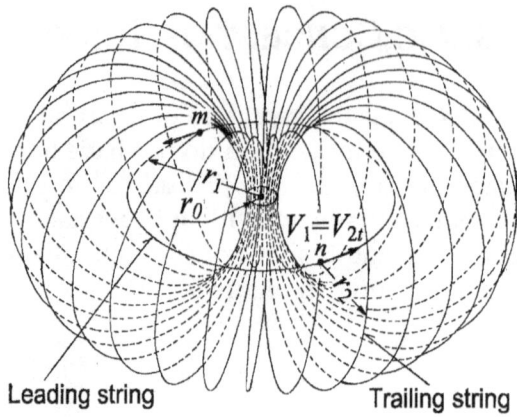

Figure 5.1.1. Isometric view of a toryx.

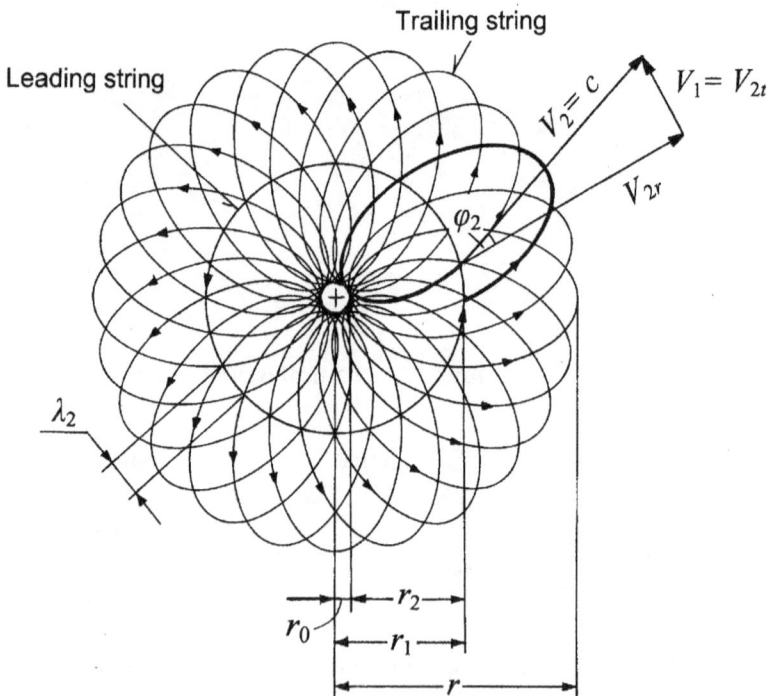

Figure 5.1.2. Top view of a toryx.

Double-toroidal trailing string – The double-toroidal trailing string is made up of two toroidal branches with one of them orbiting around the point *m* and the other one around the point *n* of the double-circular leading string. The circular opening in the toryx center with the radius r_0 is called the *toryx eye*.

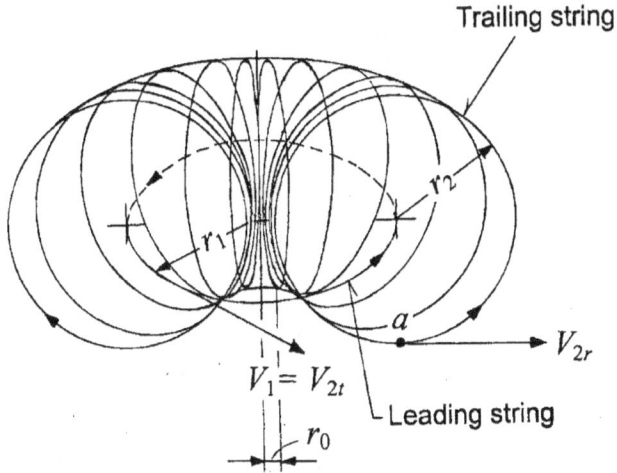

Figure 5.1.3. Cross-section of a toryx.

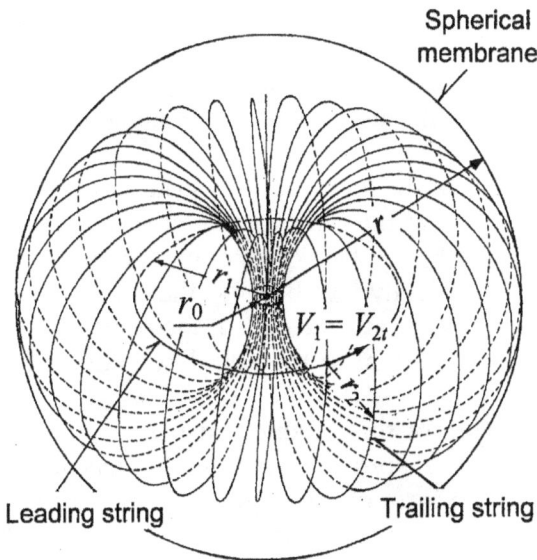

Figure 5.1.4. Complete assembly of a toryx.

Spherical membrane – Spherical membrane embraces the toryx trailing string as shown in Figure 5.1.4 and its radius is equal to the toryx peripheral radius r. The spherical membrane forms the *toryx fine structure* described in the next chapter.

5.2 Toryx Spacetime Parameters - Similarly to the helical spiral, the relationships between the spacetime parameters of the toryx are also described by the Pythagorean Theorem – see Figure 5.2.

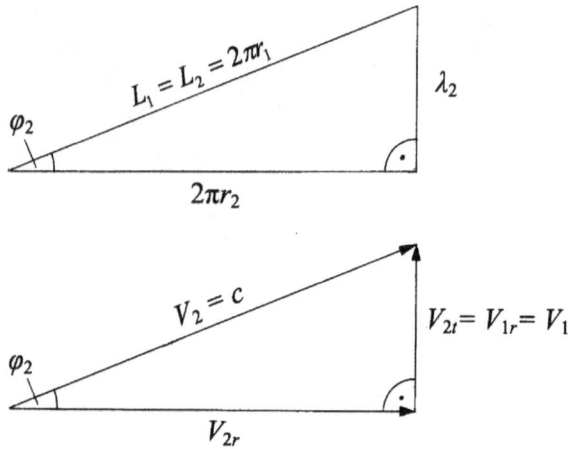

Figure 5.2. Toryx spacetime parameters expressed in absolute values corresponding to the middle point a of trailing string (Fig. 5.1.3).

Both branches of the toryx trailing string propagate along their toroidal spiral paths at the same spiral velocity V_2. In each branch, the spiral velocity V_2 has two components: rotational velocity V_{2r} and translational velocity V_{2t}. Because the propagations of trailing and leading strings are synchronized, the translational velocity of trailing string V_{2t} is equal to the spiral velocity of leading string V_1.

The application of this theorem in the toryx is more complicated than in a helical spiral. Because in the helical spiral the leading string is in the form of a straight line, the values of its spacetime parameters are independent on the position of a point of measurement located along the trailing string. In the toryx, the trailing string is bent around a circular leading string. Consequently, the values of some toryx spacetime parameters are dependent on the position of a point of measurement of these parameters. For instance, as one can see in Figure 5, the distances between windings of the toryx trailing string are shortest near the toryx eye and longest at the periphery of the trailing string.

The position-independent toryx spacetime parameters include:

f_0 = toryx base frequency
f_1 = frequency of toryx leading string
f_2 = frequency of toryx trailing string
L_1 = spiral length of one winding of toryx leading string
L_2 = spiral length of one winding of toryx trailing string
r = toryx peripheral radius, radius of spherical membrane
r_0 = toryx eye radius
r_1 = radius of toryx leading string
r_2 = radius of toryx trailing string
T_0 = toryx base period
T_1 = period of toryx leading string
T_2 = period of toryx trailing string
U_2 = volume of toryx trailing string
V_1 = spiral velocity of toryx leading string
V_{1r} = rotational velocity of toryx leading string
V_{1t} = translational velocity of toryx leading string
V_2 = spiral velocity of toryx trailing string
w_1 = the number of windings of toryx leading string
w_2 = the number of windings of toryx trailing string
λ_1 = wavelength of toryx leading string
φ_1 = steepness angle of toryx leading string.

The position-dependent toryx spacetime parameters include:

V_{2r} = rotational velocity of toryx trailing string
V_{2t} = translational velocity of toryx trailing string
λ_2 = wavelength of toryx trailing string
φ_2 = steepness angle of toryx trailing string.

5.3 "Toryx Sacred Code" (in absolute values) – The toryx sacred code is a set of three fundamental equations limiting the degrees of freedom of several toryx parameters, making possible to establish relationships between all spacetime parameters of the toryx.

Here is how the toryx sacred code looks when the toryx spacetime parameters are expressed in absolute values.

"Toryx sacred code"
(utilizing absolute values of the toryx spacetime parameters)

- The length of one winding of trailing string L_2 is equal to the length of one winding of leading string L_1:

$$L_2 = L_1 = 2\pi r_1 \quad (+\infty > r_1 > -\infty) \qquad (5.3\text{-}1)$$

- The toryx eye radius r_0 is equal to a real positive constant:

$$r_0 = r_1 - r_2 = const. \quad (-\infty < r_1 < +\infty) \qquad (5.3\text{-}2)$$

- The spiral velocity of the trailing string V_2 is constant and equals to the velocity of light in vacuum c at each point of its spiral path. Its components, the translational velocity V_{2t} and the rotational velocity V_{2r}, relate to the spiral velocity V_2 by the Pythagorean Theorem:

$$V_2 = \sqrt{V_{2t}^2 + V_{2r}^2} = c = const. \quad (-\infty < r_1 < +\infty) \qquad (5.3\text{-}3)$$

There was a good reason why I called the above three spacetime equations the "toryx sacred code." In spite of their outmost simplicity, these equations provided toryces with amazing capabilities, including:

- A capability to exist in polarized excitation and oscillation states
- A capability to form elementary matter particles capable to sustain their existence by absorption and release of energy.
- A capability to describe spacetime and physical properties of all elementary particles based on a limited set of equations.
- A capability to form elementary radiation spacetimes called *helyces* governed by the "helyx sacred code" similar to the "toryx sacred code."

- A capability of the helyces to form elementary radiation particles capable to sustain their existence by absorption and release of energy, with some of them propagating with superluminal velocity.

Amazingly, governed by similar equations, the toryces play major role in the structures of entities of both the micro- and macro-worlds.

5.4 Fundamental *Space* Property of Toryx – This property is hidden in Eqs. (5.3-1) and (5.3-2).

Let us start from Eq. (5.3-1). This equation is valid within the range of the radius of the toryx leading string r_1 extending from positive to negative infinity. Therefore, when r_1 is negative the toryx lengths L_1 and L_2 are also negative. We can easily envision a toryx in which all three parameters r_1, L_1 and L_2 are positive. In that case the toryces will appear like the ones shown in Figs. 5.1.1 – 5.1.3. Similarly, according to Eq. (5.3-2), the radius of the toryx trailing string r_2 is positive when the radius of the toryx leading string r_1 is greater than the radius of the toryx eye r_0. But, when the radius r_1 is less than r_0 the radius of trailing string r_2 becomes negative. This behavior of the toryx exposes its fundamental space property summarized below.

As the radius of the toryx leading string r_1 changes from positive to negative infinity, the toryx transforms into four completely different shapes. These transformations involve inversions (or turning inside out) of the toryx leading and trailing strings. The changes of the signs of r_1, r_2, L_1 and L_2 coincide with the changes of the toryx inversion states.

5.5 Fundamental *Spacetime* Property of Toryx – This property is hidden in Eq. (5.3-3) limiting the toryx velocities. I recognized that the innocently-looking Pythagorean Theorem allows for either translational or rotational velocity of the toryx trailing string to exceed velocity of light.

You may see it for yourself. This equation tells us that the spiral velocity of the trailing string V_2 of the toryx is constant and equals to the velocity of light c. At the same time, it says absolutely nothing about the values of its components, the translational velocity V_{2t} and the rotational velocity V_{2r}. Therefore, Equation (5.3-3) allows these velocities to be subluminal, luminal and superluminal.

The concept that the velocity of light is the maximum velocity in the universe is applied only to the spiral velocity of the toryx trailing string. But, its components, the translational and rotational velocities, can be either subluminal or superluminal.

5.6 "Toryx Sacred Code" (in relative values) - The "toryx sacred code" in absolute values can be further simplified by expressing the toryx spacetime parameters in relative values in respect to the two constant toryx parameters: the toryx eye radius r_0 and the velocity of light c.

The toryx relative spacetime parameters are:

$b = r/r_0$ = toryx relative peripheral radius
$b_1 = r_1/r_0$ = relative radius of toryx leading string
$b_2 = r_2/r_0$ = relative radius of toryx trailing string
$l_1 = L_1/2\pi r_0$ = relative length of toryx leading string
$l_2 = L_2/2\pi r_0$ = relative length of toryx trailing string
$t_1 = T_1/T_0$ = relative period of toryx leading string
$t_2 = T_2/T_0$ = relative period of toryx trailing string
$u_2 = U_2/2\pi^2 r_1^3$ = relative volume of toryx trailing string
$\beta_2 = V_2/c$ = relative spiral velocity of trailing string
$\beta_{2t} = V_{2t}/c$ = relative translational velocity of trailing string
$\beta_{2r} = V_{2r}/c$ = relative rotational velocity of trailing string
$\delta_1 = f_1/f_0$ = relative frequency of toryx leading string
$\delta_2 = f_2/f_0$ = relative frequency of toryx trailing string.

Here is how the toryx sacred code looks when the toryx space-time parameters are expressed in relative values.

"Toryx sacred code"
(utilizing relative values of the toryx spacetime parameters)

- The relative length of one winding of trailing string l_2 is equal to the relative length of one winding of leading string l_1 and the relative radius of leading string b_1:

$$l_2 = l_1 = b_1 \qquad (5.6\text{-}1)$$

- The toryx relative eye radius b_0 is equal to 1:

$$b_0 = b_1 - b_2 = 1 \qquad (5.6\text{-}2)$$

- The relative spiral velocity of trailing string β_2 is equal to 1 at each point of its spiral path. Its components, the translational velocity β_{2t} and the rotational velocity β_{2r}, relate to the relative spiral velocity β_2 by the Pythagorean Theorem:

$$\beta_2 = \sqrt{\beta_{2t}^2 + \beta_{2r}^2} = 1 \qquad (5.6\text{-}3)$$

The toryx base frequency f_0 is equal to:

$$f_0 = \frac{c}{2\pi r_0} \qquad (5.6\text{-}4)$$

The toryx base period T_0 is equal to:

$$T_0 = \frac{2\pi r_0}{c} \qquad (5.6\text{-}5)$$

Figure 5.6 shows the hodograph of the toryx relative space-time parameters.

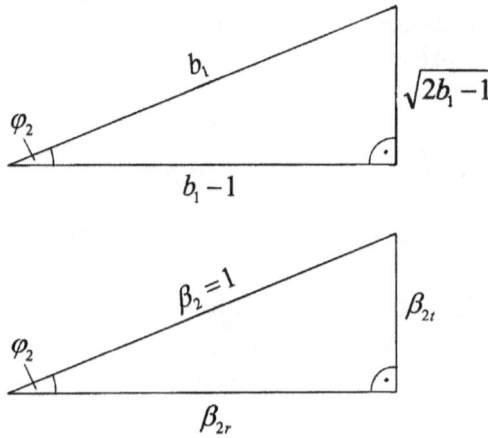

Figure 5.6. Toryx spacetime parameters expressed in relative values corresponding to the middle point a of trailing string (Fig. 5.1.3).

5.7 Toryx Derivative Spacetime Parameters - Based on the structure of the basic toryx and also on equations of the "toryx sacred code," it is possible to derive equations for all other toryx spacetime parameters shown in Tables 5.7a and 5.7b.

Table 5.7a. Relative spacetime parameters of the toryx leading and trailing strings as a function of b_1 at the middle point a of Figure 5.1.3.

Relative parameter	Leading string Eq.(a)	Trailing string Eq.(b)
Radius Eq. (5.7-1)	$b_1 = \dfrac{r_1}{r_0}$	$b_2 = \dfrac{r_2}{r_0} = b_1 - 1$
Wavelength Eq. (5.7-2)	$\eta_1 = \dfrac{\lambda_1}{2\pi r_0} = 0$	$\eta_2 = \dfrac{\lambda_2}{2\pi r_0} = \sqrt{2b_1 - 1}$
Length of one winding Eq. (5.7-3)	$l_1 = \dfrac{L_1}{2\pi r_0} = b_1$	$l_2 = \dfrac{L_2}{2\pi r_0} = b_1$
Steepness angle Eq. (5.7-4)	$\varphi_2 = 0$	$\cos u\varphi_2 = \dfrac{b_1 - 1}{b_1}$
The number of windings Eq.(5.7-5)	$w_1 = 1$	$w_2 = \dfrac{b_1}{\sqrt{2b_1 - 1}}$

Table 5.7b. Relative spacetime parameters of the toryx leading and trailing strings as a function of b_1 at the middle point a of Figure 5.1.3.

Relative parameter	Leading string Eq.(a)	Trailing string Eq.(b)
Translational velocity Eq.(5.7-6)	$$\beta_{1t} = \frac{V_{1t}}{c} = 0$$	$$\beta_{2t} = \frac{V_{2t}}{c} = \frac{\sqrt{2b_1 - 1}}{b_1}$$
Rotational velocity Eq.(5.7-7)	$$\beta_{1r} = \frac{V_{1r}}{c} = \frac{\sqrt{2b_1 - 1}}{b_1}$$	$$\beta_{2r} = \frac{V_{2r}}{c} = \frac{b_1 - 1}{b_1}$$
Spiral velocity Eq.(5.7-8)	$$\beta_1 = \frac{V_1}{c} = \frac{\sqrt{2b_1 - 1}}{b_1}$$	$$\beta_2 = \frac{V_2}{c} = 1$$
Frequency Eq. (5.7-9)	$$\delta_1 = \frac{f_1}{f_0} = \frac{\sqrt{2b_1 - 1}}{b_1^2}$$	$$\delta_2 = \frac{f_2}{f_0} = \frac{1}{b_1}$$
Period Eq.(5.7-10)	$$t_1 = \frac{T_1}{T_0} = \frac{b_1^2}{\sqrt{2b_1 - 1}}$$	$$t_2 = \frac{T_2}{T_0} = b_1$$
Volume Eq.(5.7-11)	$$u_1 = \frac{U_1}{2\pi r_0} = 0$$	$$u_3 = \frac{U_2}{2\pi^2 r_0^3} = b_1 (b_1 - 1)^2$$

Note: $\cos u\varphi_2$ is the cosine of the unified trigonometric function described in Chapter 7.

The toryx relative peripheral radius b and the relative radius of leading string b_1 are related to one another by the equation:

$$b = 2b_1 - 1 \tag{5.7-12}$$

CHAPTER 6

TORYX FINE STRUCTURE

According to our theory, toryces exist in certain excitation quantum energy states that are determined by their fine structures. One can see an obvious similarity between the excitation properties of toryces and atomic electrons described by the Bohr's model of hydrogen atom briefly described in Chapter 1.

6.1 Excited Toryces – Similarly to the atomic electrons, toryces can be either excited or de-excited. In both cases the toryx eye radius r_0 remains constant. During its excitation the radius of the toryx leading string r_1 increases while during its de-excitation this radius decreases as shown in Figure 6.1.1.

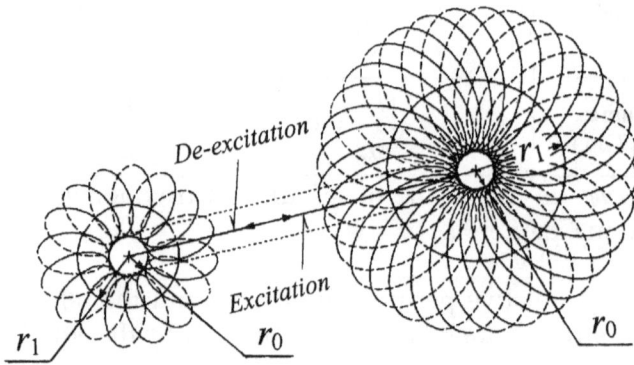

Figure 6.1.1. Excitation and de-excitation of a toryx.

The toryx can be thought as a vibrating field. We cannot hear the sounds produced by this field, because the frequencies of its vibrations are much higher than the sound frequencies of familiar to us string instruments like violins and guitars. Still, the physical principle of generation of oscillations by the toryces and the real musical string instruments is the same.

The sound waves radiating from a violin (Fig. 6.1.2) is mainly a result of interaction between its strings and body. The strings of a violin are stretched across the bridge and nut so the string ends become stationary. When a violinist draws a bow across a string, it makes the string to vibrate by the creation of standing waves. The frequency of vibration of standing waves depends on the string tension and length. Violinists shorten the playing length of the string by pushing it against the fingerboard of the violin with a finger, creating the sounds of various desired pitches. This is how the standing waves are created in the musical instruments, but do they exist in the micro-world?

Figure 6.1.2. The violin.

The first scientist who attempted to answer this question was the French physicist Louis de Broglie. While doing his Ph.D. thesis, he asked himself a logical question. If the waves behave like particles then the reverse must also be true, and the particles must also behave as waves. De Broglie generalized this statement by making a profound proposition that each particle is accompanied by a wave, which guided or "piloted" the particle through space. Moreover, it was not an electromagnetic wave but a mechanical wave, propagating faster than velocity of light. De Broglie envisioned this wave as a standing wave similar to a musical standing wave created by plucking a guitar string. He proposed that all of the possible states of the electron in the atom are standing wave states, each with its own wavelength (Fig. 6.1.3).

The electron in the atom no longer appeared as a planet in a solar system, but more like a vibrating circular string. By using his model, de Broglie was now able to provide a clearly visualizable explanation of the allowed states of the electron orbits. One simply needs to assume that the allowed states arise because the electron matter waves form standing waves, when an integral number n of wavelengths fits exactly into the circumference of a circular orbit. Thus, for the orbit with radius r we obtain that $nh = 2\pi r$. This relation leads precisely to Bohr's equation for quantization of electron orbits that we described earlier.

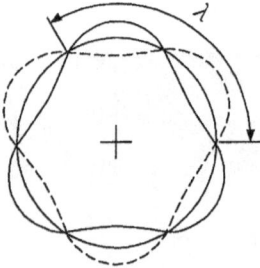

Figure 6.1.3.
Standing waves fit
to a circular Bohr orbit
of an electron for $n = 3$.
Adapted from
R.A. Serway (1990).

Two important attributes, however, are missing in the de Broglie's model of atomic electron. These drawbacks become obvious when one compares the de Broglie's model with a string instrument like a violin. Firstly, in a violin the standing waves bounce between stationary nodes. Secondly, the violin body amplifies the amplitudes of the sound frequencies produced by the oscillating violin strings. Notably, de Broglie's model contains neither the stationary nodes nor a body that resonates with the frequencies of standing waves. The UST resolves these two problems by introducing the toryx fine structure made up of two parts, the *spherical membrane* and the *standing wave*.

6.2 Spherical Membrane – The purpose of the spherical membrane is two-fold: firstly, it provides stationary nodes for the standing string and, secondly, it serves as a resonator assisting in sustaining the oscillations of the standing wave.

It was my daughter Ellen Orner who brought to my attention the idea of a possible existence of a spherical spiral in the atomic structure. Following her intuitive proposition, I came up with a solution shown in Figure 6.2.1. There a toryx is embraced by a spherical membrane, so the radius of the spherical membrane is equal to the toryx peripheral radius r.

The spherical membrane is formed by a circular motion of the membrane spiral string which windings intersect at the membrane North and South poles indicated by the letters a and b. The membrane string propagates along its spiral path at the spiral velocity V_3 equal to the velocity of light c. Thus:

$$V_3 = c \qquad (6.2\text{-}1)$$

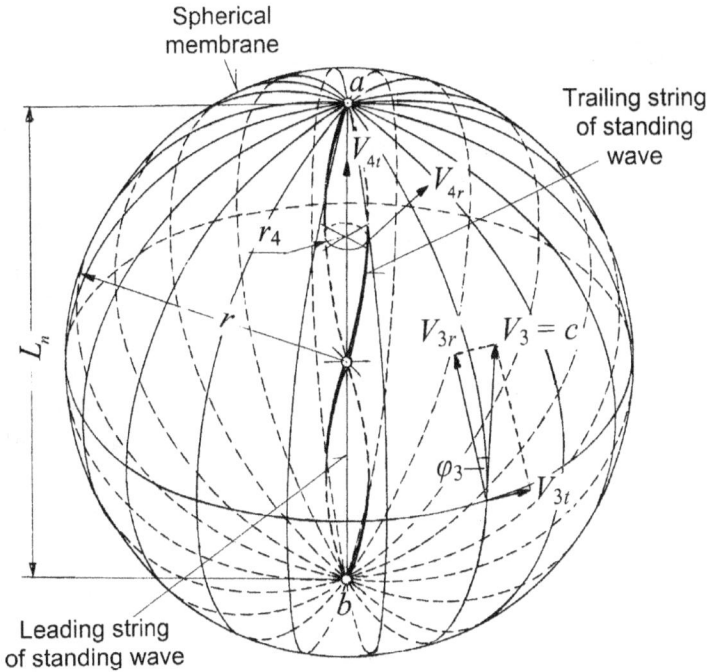

Figure 6.2.1. Spherical membrane with standing wave.

There are two components of the spiral velocity V_3 of the membrane string: the translational velocity V_{3t} and the rotational velocity V_{3r}. The relationship between the membrane velocities are defined by the Pythagorean Theorem:

$$V_3 = \sqrt{V_{3t}^2 + V_{3r}^2} = c \qquad (6.2\text{-}3)$$

For the spherical membrane to be in the resonance with the toryx, the translational frequency f_{3t} of the membrane string is equal to the frequency of the toryx leading string f_1. Thus,

$$f_{3t} = f_1 \qquad (6.2\text{-}4)$$

Table 6.2 shows equations for the spacetime parameters of spherical membrane as a function of the relative radius of toryx leading string b_1.

Table 6.2. Relative spacetime parameters of the toryx spherical membrane as a function of the relative radius of toryx leading string b_1.

Relative parameter	Translational component of velocity Eq.(a)	Rotational component of velocity Eq.(b)
Wavelength Eq. (6.2-5)	$\eta_{3t} = \dfrac{\lambda_{3t}}{2\pi r_0} = 2b_1 - 1$	$\eta_{3r} = \dfrac{\lambda_{3r}}{2\pi r_0} = \dfrac{(2b_1-1)^{3/2}}{b_1}$
Number of windings Eq.(6.2-6)	$w_{3t} = 1$	$w_{3r} = \dfrac{b_1}{\sqrt{2b_1-1}}$
Velocity Eq.(6.2-7)	$\beta_{3t} = \dfrac{V_{3t}}{c} = \dfrac{(2b_1-1)^{3/2}}{b_1^2}$	$\beta_{3r} = \dfrac{V_{3r}}{c} = \dfrac{\sqrt{b_1^4 - (2b_1-1)^3}}{b_1^2}$
Frequency Eq. (6.2-8)	$\delta_{3t} = \dfrac{f_{3t}}{f_0} = \dfrac{\sqrt{2b_1-1}}{b_1^2}$	$\delta_{3r} = \dfrac{f_{3r}}{f_0} = \dfrac{1}{b_1}$
Period Eq.(6.2-9)	$t_{3t} = \dfrac{f_0}{f_{3t}} = \dfrac{b_1^2}{\sqrt{2b_1-1}}$	$t_{3r} = \dfrac{f_0}{f_{3r}} = b_1$

6.3 Standing Wave – The standing wave oscillated between the membrane stationary nodes a and b as shown in Figure 6.2.1.

The standing string is made up of a linear leading string and a helical trailing string with the wavelength λ_4 defined by the equation:

$$\lambda_4 = \frac{2L_n}{n} \qquad (n = 0,1,2..,) \qquad (6.3\text{-}1)$$

where
L_n = distance between the membrane stationary nodes a and b
n = toryx linear excitation quantum state.

The spiral length L_4 of trailing string of standing wave can be defined by the Pythagorean Theorem:

$$L_4 = \sqrt{\lambda_4^2 + (2\pi r_4)^2} \qquad (6.3\text{-}2)$$

The aspect ratio of the standing wave is assumed to be equal to the ratio of the wavelength λ_4 to the diameter $(2r_4)$ of its trailing string and equal to the *unified quantization constant* Λ.

$$\frac{\lambda_4}{2r_4} = \Lambda = const. \qquad (6.3\text{-}3)$$

For the standing wave to be in the resonance with the toryx, the frequency of its trailing string f_4 is equal to the frequency of the toryx trailing string f_2. Thus,

$$f_4 = f_2 \qquad (6.3\text{-}4)$$

Tables 6.3a and 6.3b show equations for the spacetime parameters of standing wave as a function of the relative radius of toryx leading string b_1.

Table 6.3a. Relative spacetime parameters of trailing string of the toryx standing wave as a function of b_1.

Relative parameter	Equation	
Radius	$b_4 = \dfrac{r_4}{r_0} = \dfrac{2(2b_1 - 1)}{n\Lambda}$	(6.3-5)
Wavelength	$\eta_4 = \dfrac{\lambda_3}{r_0} = \dfrac{4(2b_1 - 1)}{n}$	(6.3-6)
Spiral length	$l_4 = \dfrac{L_4}{r_0} = \dfrac{4U(2b_1 - 1)}{n\Lambda}$	(6.3-7)
Translational velocity	$\beta_{4t} = \dfrac{V_{4t}}{c} = \dfrac{2(2b_1 - 1)^{3/2}}{\pi n b_1^2}$	(6.3-8)
Rotational velocity	$\beta_{4r} = \dfrac{V_{4r}}{c} = \dfrac{2(2b_1 - 1)^{3/2}}{n\Lambda b_1^2}$	(6.3-9)
Spiral velocity	$\beta_4 = \dfrac{V_4}{c} = \dfrac{2U(2b_1 - 1)^{3/2}}{\pi n\Lambda b_1^2}$	(6.3-10)

Table 6.3b. Relative spacetime parameters of trailing string
of the toryx standing wave as a function of b_1.

Frequency	$\delta_4 = \dfrac{f_4}{f_0} = \dfrac{\sqrt{2b_1 - 1}}{b_1^2}$	(6.3-11)
Period	$t_4 = \dfrac{f_0}{f_4} = \dfrac{b_1^4}{\sqrt{2b_1 - 1}}$	(6.3-12)

In Eq. (6.3-10), U is the *unified fine structure constant* related to
the unified quantization constant Λ by the equation:

$$U = \sqrt{\Lambda^2 + \pi^2} \qquad (6.3\text{-}13)$$

Notably, Eq. (6.3-13) is similar to the following equation proposed by T.J. Burger as an approximation of the inverse fine structure constant α^{-1}:

$$\alpha^{-1} = \sqrt{\Lambda^2 + \pi^2} \qquad (6.3\text{-}14)$$

When Λ = 137, this equation yields the value of α^{-1} =
137.036015720. The relative difference between this values and
the experimental value of α^{-1} = 137.035999074 provided by 2011
CODATA is about one part per ten million. Therefore, we may
propose that the unified fine structure constant U is equal to the
inverse fine structure constant α^{-1}, so:

$$U = \alpha^{-1} \qquad (6.3\text{-}15)$$

Incidentally, I found that the value of 137 appears in the series
shown below:

$$\Lambda_0 = 2^0 = 1$$
$$\Lambda_1 = 2^1(1^1 + 2^1) = 6$$
$$\Lambda_2 = 2^2(1^2 + 2^2) + 3^2(2^2 + 3^2) = 137$$
$$\Lambda_3 = 2^3(1^3 + 2^3) + 3^3(2^3 + 3^3) + 4^3(3^3 + 4^3) = 6841$$

....

6.4 Fine Structure Constant – Introduced by Arnold Sommerfeld, the fine structure constant α is dimensionless and
approximately equal to the inverse of 137. The fine structure

constant showed up later in many models of elementary particles proposed by other physicists.

Table 6.4 Equations for calculation of the inverse fine structure constant α^{-1}.

Author(s)	Equations	Error, %
G.N. Lewis E.Q. Adams (1914)	$\alpha^{-1} = 8\pi(8\pi^5/15)^{1/3}$	+0.22773
A.S. Eddington (1930)	$\alpha^{-1} = (16^2 - 16)/2 + 16 + 1$	-0.02627
W. Heisenberg (1935)	$\alpha^{-1} = 2^4 3^2/\pi$	+0.34580
A.M. Wyler (1971)	$\alpha^{-1} = (8\pi^4/9)(2^4 5!/\pi^5)^{1/4}$	+0.00006
B. Robertson (1971)	$\alpha^{-1} = 2^{-19/4} 3^{10/3} 5^{17/4}/\pi^2$	-0.00005
H. Aspden D.M. Eagles (1972)	$\alpha^{-1} = 108\pi(8/1843)^{1/6}$	-0.00006
T.J. Burger (1978)	$\alpha^{-1} = (137^2 + \pi^2)^{1/2}$	+0.00001
J.G. Gilson (2006)	$\alpha^{-1} = \pi/(29\cos(\pi/137)\tan(\pi/137\cdot 29))$	0.000000
V.B. Ginzburg (2006)	$\alpha^{-1} = e^{3D/e}; \quad D = \sqrt{\pi^2 + e^2 + \phi^2}$	+0.01387

Table 6.4 shows equations for calculations of the inverse fine structure constant α^{-1} proposed by several research workers. Notably, all equations, except for one proposed by Eddington contain π. The equation proposed by Ginzburg contains, besides π, two other constants, the base of natural logarithm e and the golden ratio ϕ. Table 6.4 shows the relative error between calculated values of α^{-1} and experimental values of $\alpha^{-1} = 136.035999074$ provided by 2011 CODATA.

Because the constant does not have dimensions, some scientists believe that it has a much greater significance in nature than the constants with dimensions invented by people.

CHAPTER 7

UNIFIED TORYX MATH

A notion that the toryx requires to use a special math have not come to my mind when I started to derive the spacetime equations of the toryx based on its "sacred code." As I mentioned in Chapter 5, my goal was to make these equations valid within the range of the relative radius of the toryx leading string b_1 extending from positive to negative infinity. Expecting no problems in accomplishing this goal, I began my derivations by using a commonly-known elementary math. But, very soon I realized that elementary math will make the derived equations valid only within a limited range of b_1, significantly reducing a possible range of toryx properties allowed by the "toryx sacred code."

There was actually nothing new in the fact that a curved spacetime would require using a more sophisticated math than elementary one. Albert Einstein used in his General Theory of Relativity a special branch of mathematics called the *tensor calculus*. The tensor calculus was a very new branch of math at his time and he was lucky to receive some help from his close friend Marcel Grossman.

A much needed help came to me unexpectedly from the "toryx sacred code." My analysis of its three fundamental equations revealed that it is still possible to use elementary math as a basis in application to the toryx. However, it will be necessary to introduce several modifications of elementary math to describe the inversions of the toryx spacetimes.

As I will describe later, the modifications of elementary math led me to the unification of several principal mathematical terms, explaining why I added the adjective "unified" to the title of the toryx math. Let us begin from describing a role of polarization in the unified toryx math.

7.1 Polarized Universe - Both toryces and helyces are able to exist in polarized states. There is a slight problem, however, in presenting this concept. As the peace-loving people, we often associate polarization with an undesirable confrontation.

Still, whether we like it or not, our world is made up of opposites. Moreover, take the opposites away and the world will cease to exist as it is illustrated below in my short poem.

We Live in a Polarized Universe

- There is no *young* without *old*
- There is no *hot* without *cold*.

    - There is no *high* without *low*
    - There is no *fast* without *slow*.

- There is no *dark* without *light*
- There is no *left* without *right*.

    - There is no *hiding* without *unveiling*
    - There is no *inhaling* without *exhaling*.

- There is no *weak* without *strong*
- There is no *correct* without *wrong*.

    - There is no *loosing* without *winning*
    - There is no *ending* without *beginning*.

- There is no *durable* without *brittle*
- There is no *too much* without *too little*.

    - There is no *less* without *more*
    - There is no *peace* without *war*.

- There is no *destruction* without *creation*
- There are no *facts* without *misinformation*.

    - There is no *favorable* without *adverse*
    - There is no *forward* without *reverse*.

- There is no *calm* without *fuss*
- There is no *minus* without *plus*.

    - There is no *sunrise* without *sunset*
    - There is no *joy* without *being upset*.

- There is no a *valley* without a *hill*
- There is no *imaginary* without *real*.

- There is no *hate* without *adoration*
- There is no *unification* without *polarization*.

7.2 Polarization in Elementary Math - Elementary math expresses polarization by employing positive and negative numbers.

Most of us are familiar with an elementary number line. As shown in Figure 7.2, the elementary number line is a set of real numbers *b* extending along a straight line from zero to the right side towards positive infinity (+∞)and to the left side towards negative infinity (−∞).

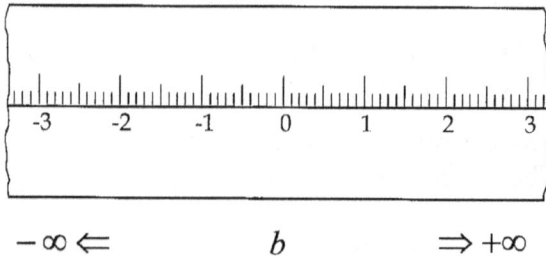

$$-\infty \Leftarrow \qquad b \qquad \Rightarrow +\infty$$

Figure 7.2. Elementary number line.

We use elementary zero as a number in two ways. Firstly, we employ it for counting of non-divisible entities in the form of an integer immediately preceding 1 in an elementary number line. Secondly, we use zero to represent an absolute absence of any quantity and quality. But the last definition of zero creates a big problem when we try to divide, for instance 1 by ever increasing numbers:

$$1 : 10 = 0.1$$
$$1 : 100 = 0.01$$
$$1 : 1000 = 0.001$$
$$1 : 10000 = 0.0001$$
$$\dots\dots\dots\dots$$
$$1 : \infty \ = 0$$

Thus, in elementary math zero (0) is mathematically defined as an inverse of infinity (∞). This, however, defies any reasonable logic. Judge yourself. After we divide the *exact number* one (1) by an *uncertain number* (∞) we will obtain the *exact number* zero (0). This also contradicts with our experience. We know that the values of natural entities, however infinitesimally small, never reach absolute zero. For instance, after removing all apples from a basket there will presumably be no, or zero, apples left in there. But this would not be true if we consider a fact that the apples are made up of zillions of molecules. Many of these molecules will still be left in the basket, giving away their characteristic smell.

This is not all. The molecules are made up of atoms, so after the removal of molecules some of their atoms will still be left in the basket. The same will happen with the nucleons containing in the atoms and with the quarks containing in the nucleons, and so on and so on. Thus, considering the sophisticated structure of apples, it would be impossible to make the basket "absolutely" empty. The above example with the apples explains why physicists are not able to reach either the "absolute zero temperature" or the "absolute vacuum." In both cases, it would require to spend an infinite amount of energy.

7.3 Infinility Versus Infinity - Elementary math expresses polarization by employing positive and negative numbers.

$$Infinility \Leftarrow \qquad\qquad \Rightarrow Infinity$$
$$(\pm 0) \qquad\qquad (\pm \ddot{1}) \qquad\qquad (\pm\infty)$$
$$Unity$$

Figure 7.3.1. Infinity $(\pm\infty)$, infinility (± 0) and unity (± 1).

Unified math clearly separates two applications of zero. The zero is still considered as an integer for counting of non-divisible entities and still retains its old symbol (0). But, in application to the real complex entities the zero is replaced with a quantity that is infinitely approaching to it. This quantity is called *infinility*, from the "infinite nil." Notably, the term infinility is similar to the known math term *infinitesimal*, but, besides to be much easier to pronounce, the term "infinility" also sounds more appropriate as a counterpart of "infinity." The role of infinility in nature is as important as the role of infinity. Figure 7.3.1 shows symbolically positive and negative infinities $(\pm\infty)$ and also positive and negative infinilities (± 0) as equal counterparts in respect to the positive and negative unities (± 1).

Unified math determines infinility in a similar same way that was used by the mathematicians to determine infinity. Most of us learned about infinity in school. Unlike the finite quantities such as 1, 2, 100, 2500, etc., an infinite quantity cannot be completely counted or measured. So, infinity is *larger* than any finite quantity. We can arrive to infinity, for instance, by multiplying a finite quantity $q = 1$ by 10, 100, 1000, etc.:

$$1\times 10 = 10^1 \qquad\qquad 1\times(-10) = -10^1$$
$$1\times 100 = 10^2 \qquad\qquad 1\times(-100) = -10^2$$
$$1\times 1000 = 10^3 \qquad\qquad 1\times(-1000) = -10^3$$
$$\cdots\cdots\cdots\cdots\cdots \qquad\qquad \cdots\cdots\cdots\cdots\cdots$$
$$\rightarrow +\infty \qquad\qquad\qquad \rightarrow -\infty$$

Infinity can be either real positive $(+\infty)$ or real negative $(-\infty)$. It can also be either imaginary positive $(+\infty i)$ or imaginary negative $(-\infty i)$ in which $i = \sqrt{-1}$.

Similarly to infinity, infinility cannot be completely counted or measured. But, unlike infinity, infinility is *smaller* than any finite quantity. We can logically arrive to infinility, for instance, by dividing 1 by 10, 100, 1000, etc.:

$$1:10 = 10^{-1} \qquad\qquad 1:(-10) = -10^{-1}$$
$$1:100 = 10^{-2} \qquad\qquad 1:(-100) = -10^{-2}$$
$$1:1000 = 10^{-3} \qquad\qquad 1:(-1000) = -10^{-3}$$
$$\cdots\cdots\cdots\cdots\cdots \qquad\qquad \cdots\cdots\cdots\cdots\cdots$$
$$\rightarrow +0 \qquad\qquad\qquad \rightarrow -0$$

Infinility can be either the real positive $(+0)$ or real negative (-0). It can also be either the imaginary positive $(+0i)$ or imaginary negative $(-0i)$. Let us summarize the above discussion about infinity and infiniliy.

Infinity $(\pm\infty)$ is a quantity that is *larger* than any given quantity.

Infinility (± 0) is a quantity that is *smaller* than any given quantity.

$$\text{Real infinility: } \pm 0 = \frac{1}{\pm \infty}; \quad \text{Imaginary infinility: } \pm 0i = \frac{1}{\pm \infty i}$$

$$\text{Real infinity: } \pm \infty = \frac{1}{\pm 0}; \quad \text{Imaginary infinity: } \pm \infty i = \frac{1}{\pm 0i}$$

The replacement of zero with infinility requires us to introduce new definitions of two other conventional geometrical terms, *point, line* and *surface*. According to the conventional geometry (Fig. 7.3.2), points are zero-dimensional, so they do not have volume, area, length, or any other higher-dimensional analogue. Lines are two-dimensional, so they have no width but can be extended on and on forever in either direction. Surfaces are three-dimensional, so they have width and length, but do not have thickness.

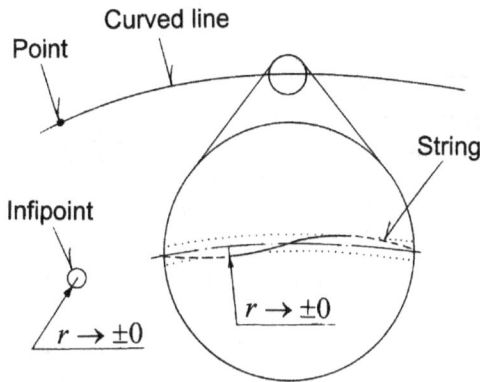

Figure 7.3.2. Infipoint and string versus point and curved line.

The UST replaces conventional definition of the point, line and surface with the respective terms called *infipoint, string* and *membrane* as summarized below:

- Infipoint is ether a circle or a sphere which radius approaches infinility (± 0).

- String is a helical spiral which radius approaches infinility (± 0).

- Membrane is a surface which thickness approaches infinil-ity (± 0).

The proposed relationship between infinity and infinility allows one to construct the *unified number line*.

7.4 Unified Number Line – As shown in Figure 7.4, unified number line is circular.

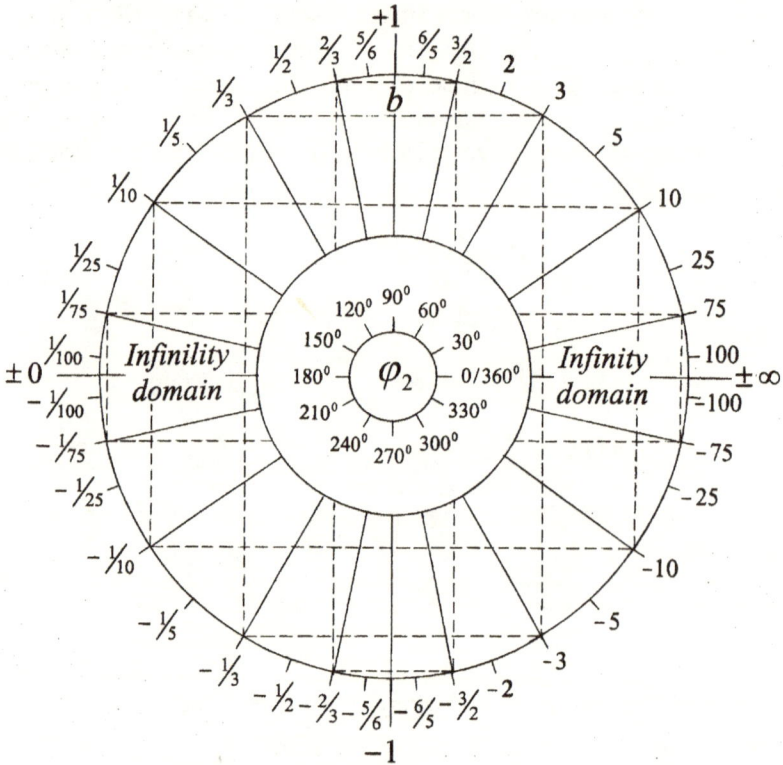

Figure 7.4. Unified number line.

In the unified number line, a set of real numbers b is extended counterclockwise along a circle from positive infinity $(+\infty)$ to negative infinity $(-\infty)$. Two equations describe the relationship between the numbers b and the polar angle φ_2 of the unified number line.

$$b = \frac{1+\cos\varphi_2}{1-\cos\varphi_2} \qquad (0^0 < \varphi_2 < 180^0) \qquad\qquad (7.4\text{-}1)$$

$$b = -\frac{1+\cos\varphi_2}{1-\cos\varphi_2} \qquad (180^0 < \varphi_2 < 360^0) \qquad\qquad (7.4\text{-}2)$$

The unified number line is divided into two domains, *infinity domain* and *infinility domain*, which occupy equal spaces on the number line.

Infinity domain contains the numbers b extending clockwise from the positive unity $(+1)$ to the negative unity (-1). It resides in two right quadrants of the number line:

<u>Top right quadrant</u> $(+0^0 < \varphi_2 < 90^0)$ - Within this quadrant the numbers b increase from the positive unity $(+1)$ to the positive infinity $(+\infty)$.

<u>Bottom right quadrant</u> $(270^0 < \varphi_2 < 360^0)$ - Within this quadrant the numbers b decrease from the negative unity (-1) to the negative infinity $(-\infty)$.

Infinility domain contains the numbers b extending counterclockwise from the positive unity $(+1)$ to the negative unity (-1). It resides in two left quadrants of the number line:

<u>Top left quadrant</u> $(90^0 < \varphi_2 < 180^0)$ - Within this quadrant the numbers b decrease from the positive unity $(+1)$ to the positive infinility $(+0)$.

<u>Bottom left quadrant</u> $(180^0 < \varphi_2 < 270^0)$ - Within this quadrant the numbers b decrease from the negative infinility (-0) to the negative unity (-1).

Notably, the negative infinity $(-\infty)$ merges with the positive infinity $(+\infty)$ at $\varphi_2 \to 0/360^0$, while the positive infinility $(+0)$ merges with the negative infinility (-0) at $\varphi_2 \to 180^0$.

There are two kinds of polarization between the numbers b of unified number line, *inverse* and *reverse*:

Inverse polarization - The magnitudes of the numbers b'' located in the left quadrants are inversed in respect to the magnitudes of the numbers b' located in the right quadrants. Thus,

$$b'' = \frac{1}{b'}$$

Reverse polarization - The numbers located in the top quadrants b' and the bottom quadrants b'' have same magnitudes but reversed signs. Thus,

$$b'' = -b'$$

The construction of unified number line requires us to replace elementary trigonometry with unified trigonometry.

7.5 Pythagorean Theorem – The Pythagorean Theorem is a at the heart of our theory. In fact it is used to express the third equation of the "toryx sacred code." A reason why we shall to re-examine the 2500-year-old Pythagorean Theorem again is because it is still not understood completely by a vast majority of young people after finishing high school.

The original concept of the Pythagorean Theorem was invented by the Babylonians who use a rope to make a perfect right triangle (Fig. 7.5). Mark twelve equally distant spaces on the rope with the knots. Divide the rope into three sections: a section 3 with three spaces, a section 4 with four spaces and section 5 with five spaces. Connect two free ends of the rope together. Stretch out section 4 and attach its ends to a flat surface. Grab the knot between sections 3 and 5 and pull both these sections away from section 4 until the entire rope is completely stretched out. The rope now will be in a form of a triangle having a perfect right angle between sections 3 and 4. This type of triangle became known as a *right triangle*.

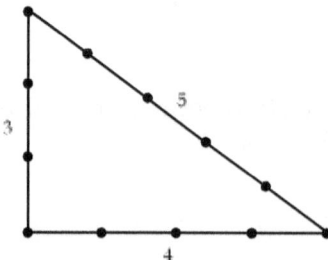

Figure 7.5.
Making a right
triangle
with a rope.

Besides the above example of the right triangle with the sides related by the ratios 3:4:5, the Babylonians were familiar with fourteen more right triangles, including those with the sides related by the ratios 5:12:13, 8:15:17, and 7:24:25. Notably, the Babylonians discovered that the square of the longest side of the right triangle is equal to the sum of the squares of the two remaining sides as shown below:

$$3^2 + 4^2 = 5^2 \ (9 + 16 = 25),$$
$$5^2 + 12^2 = 13^2 \ (25 + 144 = 169),$$
$$8^2 + 15^2 = 17^2 \ (64 + 225 = 289),$$
$$7^2 + 24^2 = 25^2 \ (49 + 576 = 625), \text{ etc.}$$

Thus, the Babylonians came very close to the formulation of what became known to us as the *Pythagorean Theorem*. The Greek philosopher Pythagoras of Samos provided a more detailed examination of the relations between sides of right triangles and summarized it in his famous Pythagorean Theorem:

In any right triangle, the sum of the squares of the two sides A and B is equal to the square of the hypotenuse C, so:

$$A^2 + B^2 = C^2 \tag{7.5-1}$$

After reading this theorem, one can easily become confused. In our previous discussions we mentioned that both the Egyptians and the Babylonians discovered many right triangles, to which this rule was applied. So, what was different in the Pythagorean Theorem? Actually, the difference was only in one word "any." Pythagoras stated that the rule, applicable to the right triangles known before his time, was also true for any right triangle. By making this theoretical generalization of practical geometry, Pythagoras made another important step towards conversion of practical geometry into science.

The impact of the Pythagorean Theorem on the development of science was much greater than its author could have ever envisioned. During 2500 years since its discovery, this theorem has proved to be invaluable for solving various problems in all branches of science, and also in engineering and art. Ironically, during his time, Pythagoras became the first scientist who tried to undermine unique versatility of his own theorem. As a true believer in the harmony of the universe, Pythagoras thought that the sides of the right triangles must always be expressed by either whole or as the ratios of whole numbers called rational numbers such as $\frac{1}{2}, \frac{1}{3}, \frac{1}{8}, ...$ He turned to be wrong.

Initially, a problem had arisen when Pythagorean attempted to express with a whole number the hypotenuse of a *right isosceles triangle* having equal short sides. Suppose each side had a length of 1 unit, then, according to the Pythagorean Theorem, the square of these sides will be equal to the hypotenuse. Therefore, $1^2 + 1^2 = 2$, and 2 is the square of the hypotenuse. Consequently, the hypotenuse is the square root of 2, or $\sqrt{2}$. After numerous attempts the Pythagoreans failed to present $\sqrt{2}$ as a ratio of whole numbers.

Horrified by their unspeakable discovery, the Pythagoreans called $\sqrt{2}$ an *irrational number*. The entire Pythagorean system of a universe, based on whole numbers, was collapsing. But, instead of facing the problem and correcting his system, Pythagoras chose to keep this discovery secret from the outside world. This decision was crucial for the future of the Pythagorean secret society which by that time began to crumble internally and externally. Inside the Pythagorean society, it led to the breakdown in mystical moral among its members.

Eventually, the inevitable had happened. One day Hippasus of Metapontum, a mathematician and the member of the society, let the secret out and set himself up as a public teacher of geometry. The punishment was swift and resolute. According to one story, Hippasus was thrown overboard a ship by some of the followers of Pythagoras; another story attributes this insidious crime to Pythagoras himself.

This story about the Pythagorean Theorem has a direct relation to the subject of our theory. Let us express the sides of a right triangle by Eq. (7.5-1) in relative terms:

$$\beta_a = \frac{A}{C}; \quad \beta_b = \frac{B}{C}; \tag{7.5-2}$$

Then we obtain from Eqs. (7.5-1) and (7.5-2):

$$\sqrt{\beta_a^2 + \beta_b^2} = 1 \tag{7.5-3}$$

Let us compare Eq. (7.5-3) with Eq. (5.6-3) of the "toryx sacred code" in which the toryx parameters are expressed in relative values:

$$\sqrt{\beta_{2t}^2 + \beta_{2r}^2} = 1 \qquad (7.5\text{-}4)$$

Mathematically, Eqs. (7.5-3) and (7.5-4) are identical. Unlike Pythagoras, we do not limit the values of the sides of the right traingle to the rational numbers. Nowadays we know these sides can be expressed by all kinds of numbers, including:

- Rational
- Irrational, including transcendental
- Positive
- Negative
- Real
- Imaginary.

The sides of the right triangle can also be either shoorter or longer than hypotenuse. For instance, when the relative rotational velocity of the toryx trailing string $\beta_{2t} = 2$, it exceeds the speed of light by a factor of two. Consequently, we find from Eq. (7.5-4) that the relative translational velocity of the toryx trailing string β_{2r} is equal to:

$$\beta_{2r} = \sqrt{1 - \beta_{2t}^2} = \sqrt{1 - 2^2} = \sqrt{-3} = 3i$$

Thus, when the translational velocity of the toryx trailing string exceeds the speed of light, the rotational velocity of the toryx trailing string is expressed by the imaginary number.

7.6 Elementary Trigonometry - Elementary trigonometry is based on the relationships between the sides of a right triangle. The definitions of elementary trigonometric functions are based on the transformations of the right triangle as a function of the non-right angle φ_2 (Fig. 7.6.1):

$$\cos\varphi_2 = x; \quad \sin\varphi_2 = y; \quad \tan\varphi_2 = y/x$$

Figure 7.6.1 shows the relationship between the main trigonometric functions and the lengths of the sides x and y for the case when the length of the hypotenuse is equal to 1. The main features of these transformations are:

- When the length of the hypotenuse of the triangles is equal to 1, the ranges of the lengths of its sides x and y are between 1 and -1.
- The triangles located at the two left quadrants are mirror images of the triangles located at the two right quadrants.
- The triangles located at the two bottom quadrants are mirror images of the triangles located at the two top quadrants.

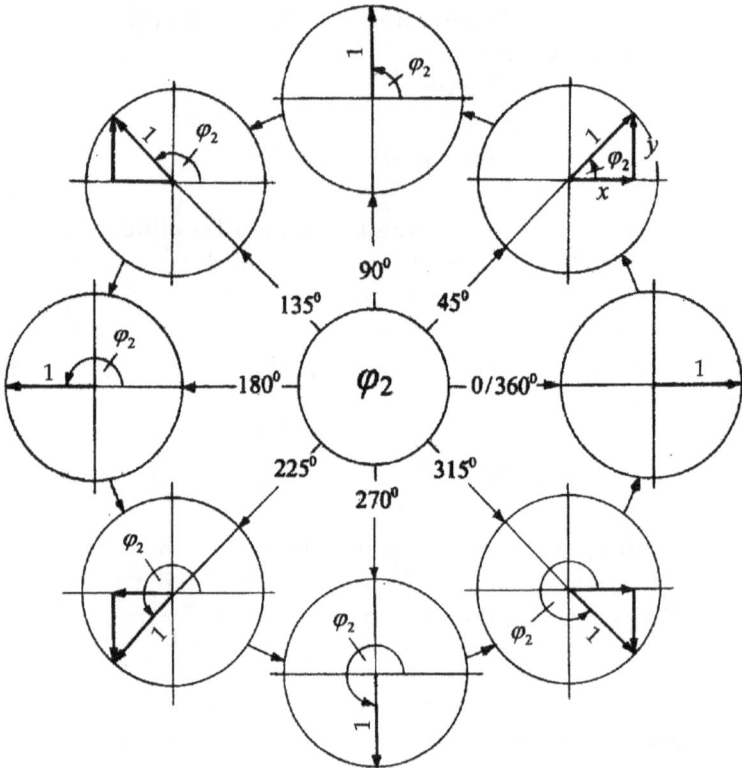

Figure 7.6.1. Transformations of a right triangle corresponding to elementary trigonometry.

7.7 Unified Trigonometry - Unified trigonometry employs the following relationship between the toryx trigonometric function $\cos u\varphi_2$ and the number b_1 that extends from positive infinity $(-\infty)$ to negative infinity $(+\infty)$ as the angle φ_2 changes from 0 to 360^0:

$$\cos u\varphi_2 = x = \frac{b_1 - 1}{b_1} \quad (0^0 < \varphi_2 < 360^0) \quad (7.7\text{-}1)$$

The letter "u" in the symbols of unified trigonometric functions differentiates them from elementary ones.

Figure 7.7 shows the transformations of the right triangle as a function of the angle φ_2 according to Eq. (7.7-1).

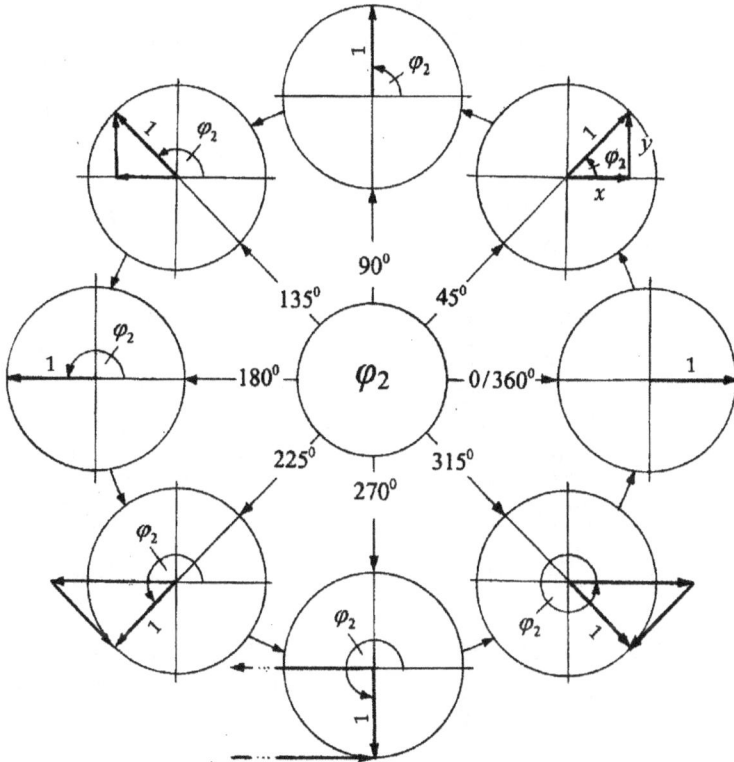

Figure 7.7. Transformations of a right triangle corresponding to unified trigonometry.

- When the angle φ_2 is between 0 and 180⁰, the right triangles in unified trigonometry look the same as in elementary trigonometry. Thus, within this range of the angle φ_2 elementary and unified trigonometry are identical.
- When the angle φ_2 is between 180 and 360⁰, the right triangle of unified trigonometry becomes *outverted*. Consequently, the length of its horizontal side x becomes greater than 1, while

the length of the other side y is expressed with imaginary numbers.

- When the angleφ_2 approaches 270^0 from a smaller angle, the length of its horizontal side x approaches positive real infinity$(+\infty)$, while the length of the other side y approaches positive imaginary infinity$(+\infty i)$.
- When the angle φ_2 approaches 270^0 from a greater angle, the length of its horizontal side x approaches negative real infinity$(-\infty)$, while the length of the other side y approaches negative imaginary infinity$(-\infty i)$.
- When the angleφ_2 approaches 360^0 from the angle smaller than 360^0, the length of its horizontal side x approaches 1, while the length of the other imaginary side y approaches zero.

Table 7.7 shows the relationship between the unified and elementary trigonometric functions.

Table 7.7 Relationship between unified and elementary trigonometric functions in application to the toryx.

Unified trigonometry	Elementary trigonometry	
$(0^0 < \varphi_2 < 360^0)$	$(0^0 < \varphi_2 < 180^0)$	$(180^0 < \varphi_2 < 360^0)$
$\cos u\varphi_2$	$\cos \varphi_2$	$\sec \varphi_2 = 1/\cos\varphi_2$
$\sin u\varphi_2$	$\sin u\varphi_2$	$-i\tan \varphi_2$

7.8 Application of Unified Toryx Math - Unified toryx math describes properly the toryx spacetime parameters when the two conditions are met:

1. The relative radius of spherical membrane b corresponds to the number b of the unified number line shown in Figure 7.4.
2. The steepness angle of the toryx trailing stringφ_2 corresponds to the polar angleφ_2 of the unified trigonometric functions and the unified number line.

Figure 7.8 shows transformations of velocities of the toryx trailing string corresponding to the middle point of trailing string as its steepness angle φ_2 increases from 0 to 360^0. The translational and rotational velocities of the toryx trailing string are expressed in relative terms in respect to the speed of light c. In each right triangles of velocities of trailing string one side represent the relative translational velocity β_{2t} and the other side the relative rotational velocity β_{2r}, while its hypotenuse represents the relative spiral velocity $\beta_2 = 1$. Obviously, there is a similarity between the transformations of the toryx velocities shown in Figure 7.8 and the transformations of a right triangle corresponding to the unified trigonometry shown in Figure 7.7.

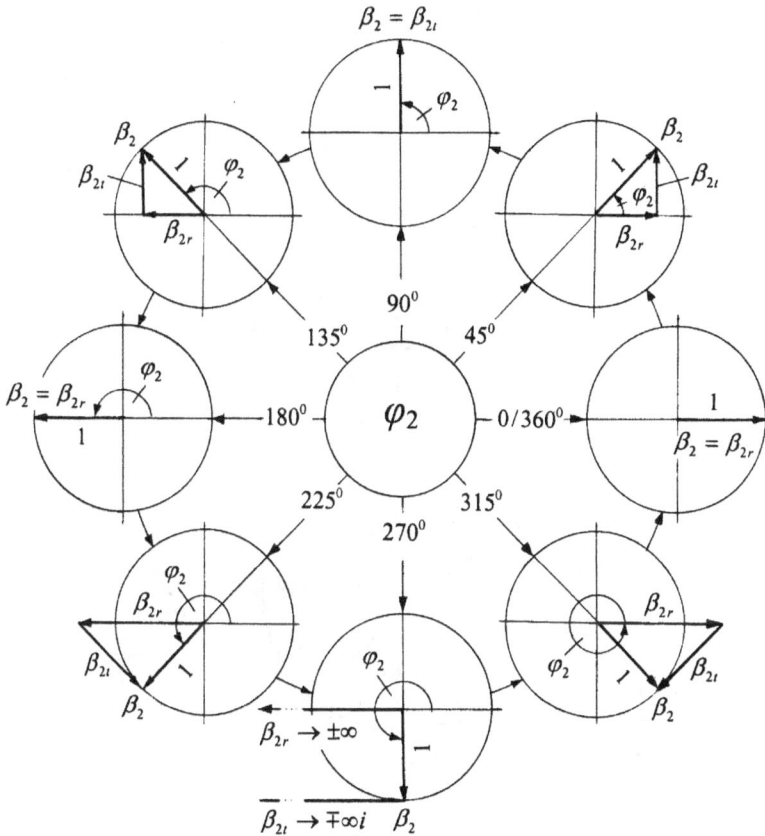

Figure 7.8. Transformations of velocities of the toryx trailing string at the middle point of trailing string.

CHAPTER 8

TRENDS OF TORYX SPACETIME PARAMETERS

Two integrated space-time parameters of toryces define their main classification: the *toryx reality* R and the *toryx vorticity* V.

8.1 Toryx Reality - The toryx reality R is equal to the toryx relative peripheral radius and to the relative radius of the toryx spherical brane b.

$$R = b = 2b_1 - 1 = \frac{1 + \cos u\varphi_2}{1 - \cos u\varphi_2} \qquad (8.1\text{-}1)$$

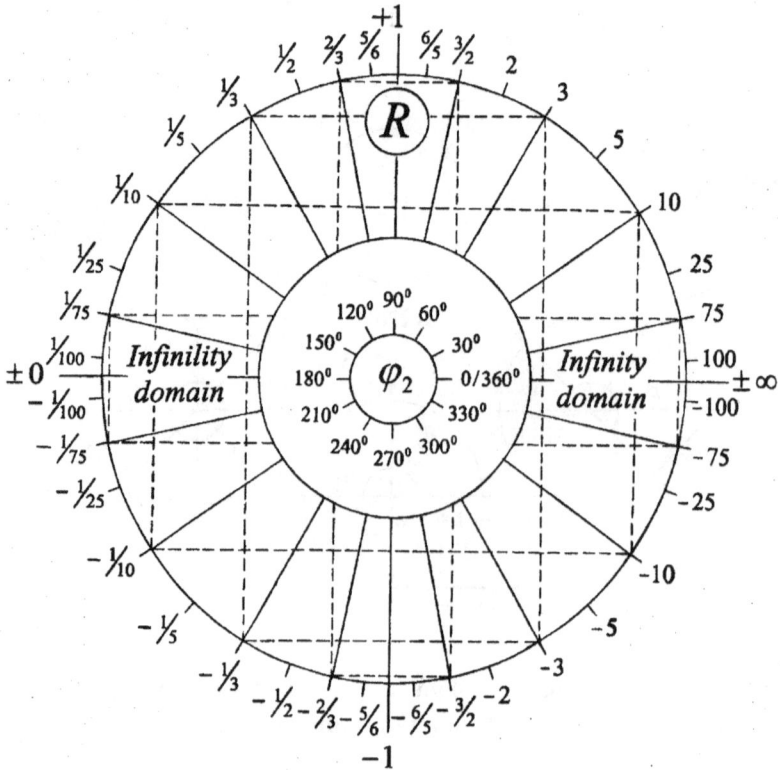

Figure 8.1.1. Toryx reality R as a function of the steepness angle of trailing string φ_2.

Figure 8.1.1 shows a circular diagram of the toryx reality R as a function of the steepness angle of trailing string φ_2. This diagram can be treated as the universal number line for the number b. The circular diagram of the toryx reality R is divided into infinility and infinity domains.

- <u>Infinility domain</u> occupies two left quadrants; it contains the values of R extending counterclockwise from the positive unity $(+1)$ and passing through infinility (± 0) to the negative unity (-1).

- <u>Infinity domain</u> resides in two right quadrants; it contains the values of R extending clockwise from the positive unity $(+1)$ and passing through infinity $(\pm\infty)$ to the negative unity (-1).

Notably, positive infinility $(+0)$ merges with negative infinility (-0) at $\varphi_2 \to 180^0$, while negative infinity $(-\infty)$ merges with positive infinity $(+\infty)$ at $\varphi_2 \to 360^0$. There are two kinds of reality polarization between the toryx realities R of toryces that belong to the four quadrants of the circular diagram, *inverse-reality* and the *reverse-reality*.

Inverse-reality polarization - The toryces that belong to the left and right quadrants are inverse-reality polarized by their reality R, because the magnitudes of their reality R are inversed while the signs of R are the same.

Reverse-reality polarization - The toryces that belong to the top and bottom quadrants are reverse-reality polarized by their reality R, because the signs of their reality R, are reversed while the magnitudes of R are the same.

8.2 Toryx Vorticity - The toryx vorticity V is equal to the ratio of the radius of trailing string r_2 to the radius of leading string r_1 with the opposite sign.

$$V = -\frac{r_2}{r_1} = -\frac{b_2}{b_1} = -\frac{b_1 - 1}{b_1} = -\frac{b - 1}{b + 1} = -\cos u\varphi_2 \quad (8.2\text{-}1)$$

Figure 8.2 shows a circular diagram of the toryx vorticity V as a function of the steepness angle of trailing string φ_2. Toryces with positive vorticity V are called *positive* and with negative vorticity V, *negative*. This diagram can be treated as a universal number line for the real numbers V.

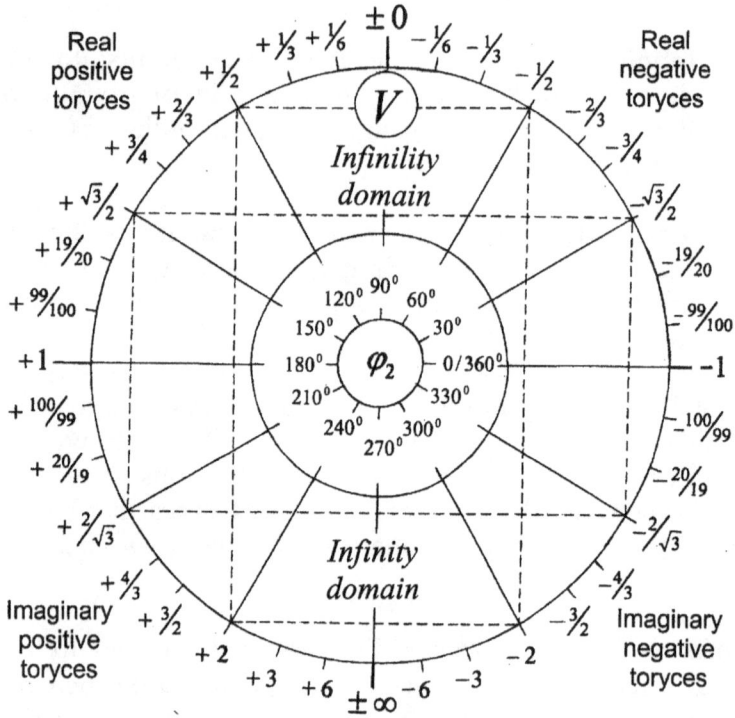

Figure 8.2. Toryx vorticity V as a function of the steepness angle of trailing string φ_2.

The circular diagram of the toryx vorticity V is divided into two domains, the *infinity domain* and the *infinility domain*, each occupying equal sectors of the diagram.

- Infinility domain occupies two top quadrants; it contains the values of V extending clockwise from the positive unity $(+1)$ and passing through infinility (± 0) to the negative unity (-1).

- Infinity domain resides in two bottom quadrants; it contains the values of V extending counterclockwise from the positive unity $(+1)$ and passing through infinity $(\pm\infty)$ to the negative unity (-1).

Notably, positive infinility $(+0)$ merges with negative infinility (-0) at $\varphi_2 \to 90^0$, while negative infinity $(-\infty)$ merges with positive infinity $(+\infty)$ at $\varphi_2 \to 270^0$. There are two kinds of vorticity polarization between the toryx vorticities V of the toryces that belong to the four quadrants of the circular diagram, *inverse-vorticity* and the *reverse-vorticity*.

Inverse-vorticity polarization - The toryces that belong to the top and bottom quadrants are inversely polarized by their vorticity V, because the magnitudes of their vorticities V are inversed while the signs of V are the same.

Reverse-vorticity polarization - The toryces that belong to the right and left quadrants are reversely polarized by their vorticity V, because the signs of their vorticities V are reversed while the magnitudes of V are the same.

8.3 Main Classification of Toryces - The toryces are divided into four main groups according to their vorticity V and reality R as shown in Table 8.3.

The sign of the vorticity V defines the sign of the toryces. In the real toryces R is positive, while in the imaginary toryces R is negative. In the real toryces all its spacetime parameters corresponding to te middle point a of trailing string (Fig. 5.1.3) are expressed with real numbers, while in the imaginary toryces several spacetime parameters are expressed with imaginary numbers.

Table 8.3. Main classification of toryces.

Toryx group	Toryx name	φ_2	V	R
1	Real negative	$0^0 - 90^0$	$(-)$	$(+)$
2	Real positive	$90^0 - 180^0$	$(+)$	$(+)$
3	Imaginary positive	$180^0 - 270^0$	$(+)$	$(-)$
4	Imaginary negative	$270^0 - 360^0$	$(-)$	$(-)$

We show below the trends of the toryx spacetime parameters as a function of the steepness angle of its trailing string φ_2. In the graphs, the real and imaginary parameters are depicted with thick and thin lines respectively. Also shown for each graph are the values of the toryx spacetime parameters corresponding to the steepness angles $\varphi_2 = 0^0$, 90^0, 180^0, 270^0 and 360^0.

8.4 Radius of Spherical Membrane - The relative radius of spherical membrane b is given by the equation:

$$b = \frac{1 + \cos u\varphi_2}{1 - \cos u\varphi_2} \qquad (0^0 < \varphi_2 < 360^0)$$

Figure 8.4 shows a plot of the above equation.

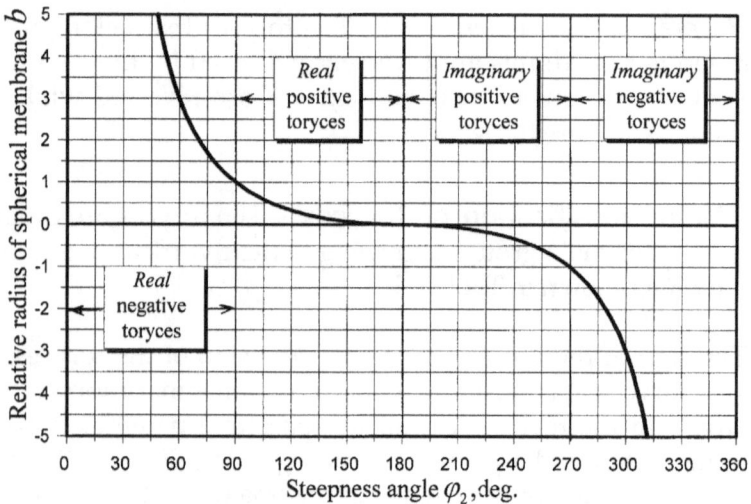

Figure 8.4. Relative radius of spherical brane b as a function of the steepness angle of trailing string φ_2.

The values of the radius of spherical membrane b corresponding to the toryx four major transformation points are shown below.

φ_2	$360/0^0$	90^0	180^0	270^0
b	$-\infty/+\infty$	$+1$	$+0/-0$	-1

8.5 Radii of Leading & Trailing Strings

- The relative radii of leading string b_1 and trailing string b_2 are given by the equations:

$$b_1 = \frac{1}{1 - \cos u\varphi_2} \qquad (0^\circ < \varphi_2 < 360^\circ)$$

$$b_2 = \frac{\cos u\varphi_2}{1 - \cos u\varphi_2} \qquad (0^\circ < \varphi_2 < 360^\circ)$$

Figure 8.5 shows a plot of the above equations.

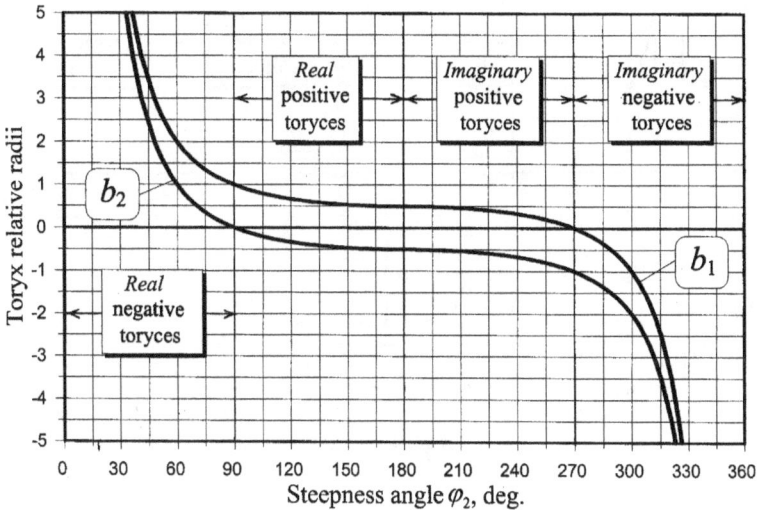

Figure 8.5. Relative radii of leading and trailing strings, b_1 and b_2, as a function of the steepness angle of trailing string φ_2.

The values of the relative radii of leading string b_1 and trailing string b_2 corresponding to the toryx four major transformation points are shown below.

φ_2	$360/0^\circ$	90°	180°	270°
b_1	$-\infty/+\infty$	$+1$	$+\frac{1}{2}$	$+0/-0$
b_2	$-\infty/+\infty$	$+0/-0$	$-\frac{1}{2}$	-1

8.6 Wavelength of Trailing String

8.6 Wavelength of Trailing String - The relative wavelength of trailing string η_2 is given by the equation:

$$\eta_2 = \sqrt{\frac{1+\cos u\varphi_2}{1-\cos u\varphi_2}} \quad (0^0 < \varphi_2 < 360^0)$$

Figure 8.6 shows a plot of the above equation.

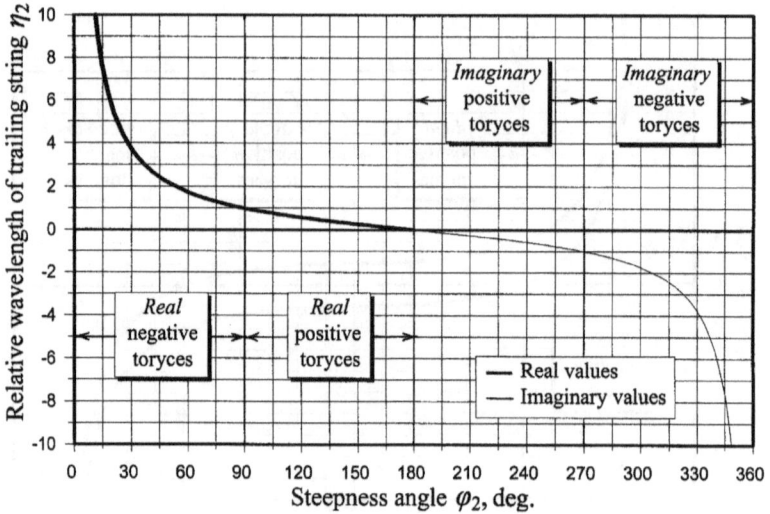

Figure 8.6. Relative wavelength of trailing string η_2 as a function of the steepness angle of trailing string φ_2.

The values of the relative wavelength of trailing string η_2 corresponding to the toryx four major transformation points are shown below.

φ_2	$360/0^0$	90^0	180^0	270^0
η_2	$-\infty i/+\infty$	$+1$	$+0/-0i$	$-i$

8.7 The Number of Windings of Trailing String - The number of windings of trailing string w_2 is given by the equation:

$$w_2 = \frac{1}{\sin u\varphi_2} \quad (0^0 < \varphi_2 < 360^0)$$

Figure 8.7 shows a plot of the above equation.

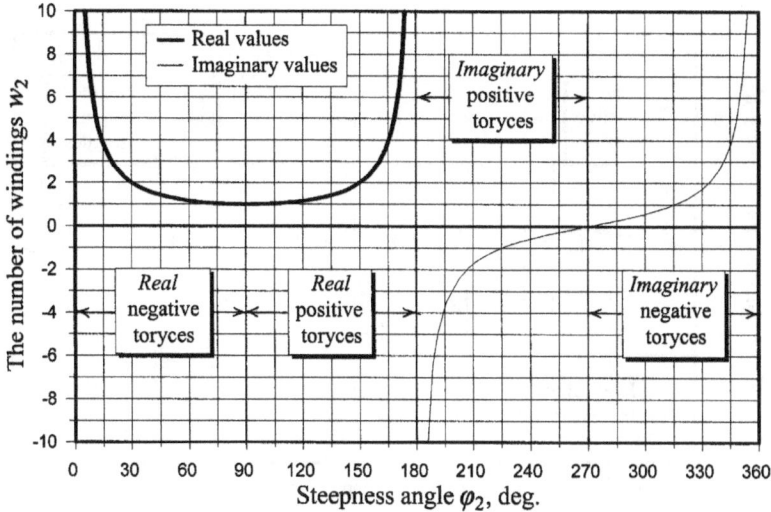

Figure 8.7. The number of windings of the trailing string w_2 as a function of the steepness angle of trailing string φ_2.

The values of the number of windings of trailing string w_2 corresponding to the toryx four major transformation points are shown below.

φ_2	$360/0^0$	90^0	180^0	270^0
w_2	$+\infty i/+\infty$	$+1$	$+\infty/-\infty i$	$-0i/+0i$

8.8 Translational Velocity of Trailing String - The relative translational velocity of trailing string β_{2t} is given by the equation:

$$\beta_{2t} = \sin u\varphi_2 \quad (0^0 < \varphi_2 < 360^0)$$

Figure 8.8 shows a plot of the above equation.

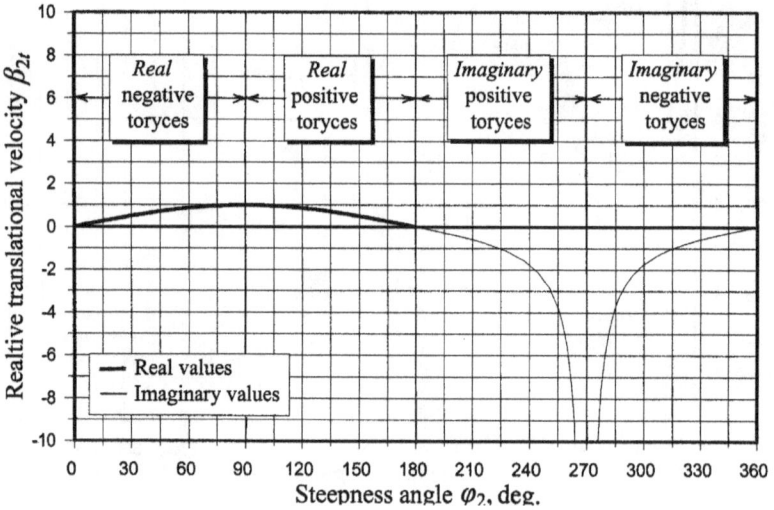

Figure 8.8. Relative translational velocity of the trailing string β_{2t} as a function of the steepness angle of trailing string φ_2.

The values of the relative translational velocity of trailing string β_{2t} corresponding to the toryx four major transformation points are shown below.

φ_2	$360/0^0$	90^0	180^0	270^0
β_{2t}	$-0i/+0$	$+1$	$+0/-0i$	$-\infty i$

8.9 Rotational Velocity of Trailing String - The relative rotational velocity of trailing string β_{2r} is given by the equation:

$$\beta_{2r} = \cos u\varphi_2 \qquad (0^0 < \varphi_2 < 360^0)$$

Figure 8.9 shows a plot of the above equation.

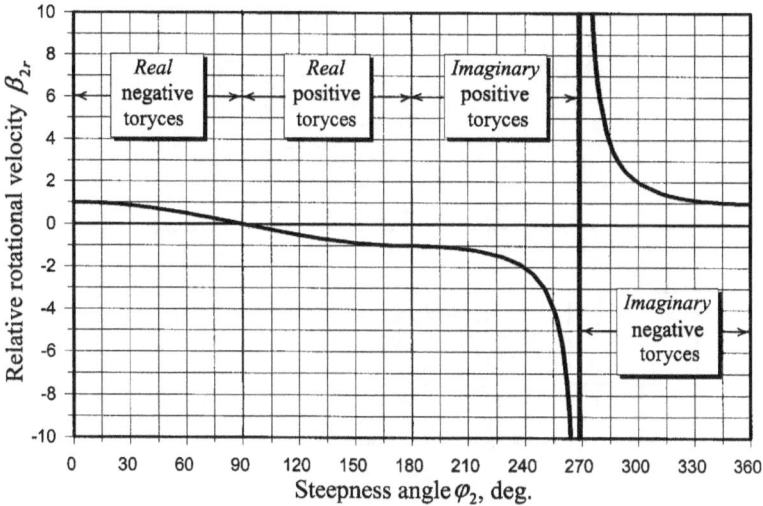

Figure 8.9. Relative rotational velocity of the trailing string β_{2r} as a function of the steepness angle of trailing string φ_2.

The values of the relative rotational velocity of trailing string β_{2r} corresponding to the toryx four major transformation points are shown below.

φ_2	$360/0^0$	90^0	180^0	270^0
β_{2r}	$+1$	$+0/-0$	-1	$-\infty/+\infty$

8.10 Frequency of Leading String - The relative frequency of leading string δ_1 is given by the equation:

$$\delta_1 = \sin u\varphi_2 (1 - \cos u\varphi_2) \quad (0^\circ < \varphi_2 < 360^\circ)$$

Figure 8.10 shows a plot the above equation.

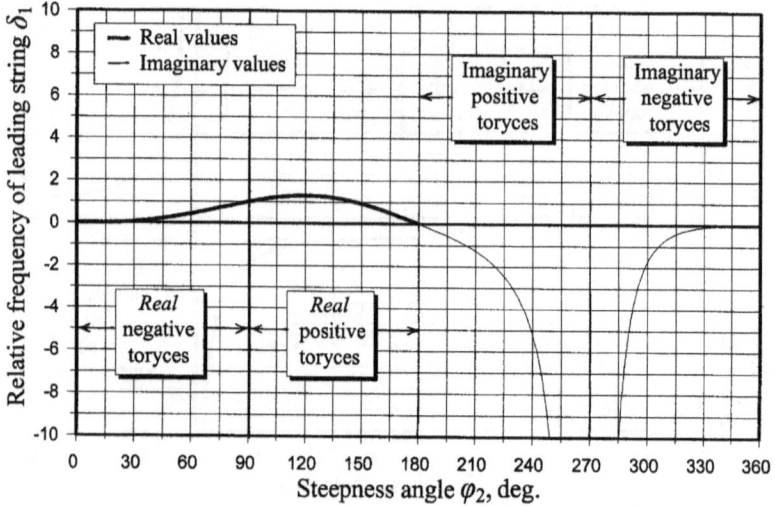

Figure 8.10. Relative frequency of leading string δ_1 as a function of the steepness angle of trailing string φ_2.

The values of the relative frequency of leading string δ_1 corresponding to the toryx four major transformation points are shown below.

φ_2	$360/0^0$	90^0	180^0	270^0
δ_1	$-0i/+0$	$+1$	$+0/-0i$	$-\infty i$

8.11 Frequency of Trailing String - The relative frequency of
trailing string δ_2 is given by the equation:

$$\delta_2 = 1 - \cos u\varphi_2 \quad (0^0 < \varphi_2 < 360^0)$$

Figure 8.11 shows a plot of the above equation.

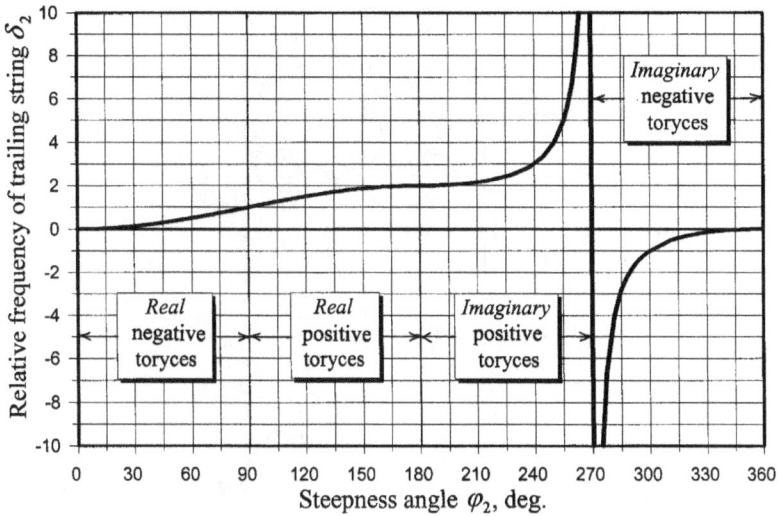

Figure 8.11. Relative frequency of trailing string δ_2 as a function of the steepness angle of trailing string φ_2.

The relative frequency of trailing string δ_2 corresponding to its four major transformation points are shown below.

φ_2	$360/0^0$	90^0	180^0	270^0
δ_2	$-0/+0$	$+1$	$+2$	$+\infty/-\infty$

8.12 Instant Velocities of Trailing String - To maintain
space-time integrity of the toryx, as its trailing string
propagates at a constant relative spiral velocity $\beta_2 = 1$, the
relative translational and rotational velocities, β_{2t} and β_{2r}
must vary in a very specific way.

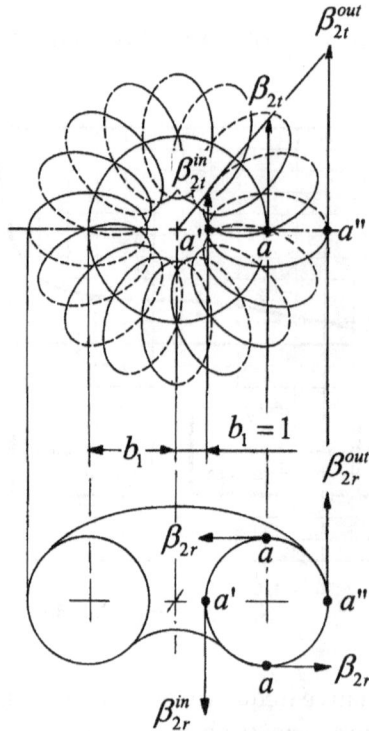

Figure 8.12.1. Relative peripheral velocities of trailing string.

As shown in Fig. 8.12.1, the translational velocity at each point
of the trailing string must be proportional to the distance of said
point from the toryx center. At the same time, the rotational velocity of the trailing string must vary to satisfy the fundamental
equation (5.6-3), according to which the trailing string must
propagate at a constant relative spiral string velocity $\beta_2 = 1$.

Table 8.12 shows equations for relative peripheral translational and rotational velocities of the trailing string. Within the
ranges of b_1 (8.12-5) and (8.12-6), the inner and outer peripheral
translational velocities β_{2t}^{in} and β_{2t}^{out} of the trailing string exceed the
velocity of light.

Table 8.12. Relative peripheral velocities of the trailing string.

Velocities at the point a'		Velocities at the point a''	
$\beta_{2t}^{in} = \dfrac{\sqrt{2b_1 - 1}}{b_1^2}$	(8.12-1)	$\beta_{2t}^{out} = \dfrac{(2b_1 - 1)^{3/2}}{b_1^2}$	(8.12-2)
$\beta_{2r}^{in} = \dfrac{\sqrt{b_1^4 - 2b_1 + 1}}{b_1^2}$	(8.12-3)	$\beta_{2r}^{out} = \dfrac{\sqrt{b_1^4 - (2b_1 - 1)^3}}{b_1^2}$	(8.12-4)
$0.544 < b_1 < 1.0$	(8.12-5)	$1.0 < b_1 < 6.222$	(8.12-6)

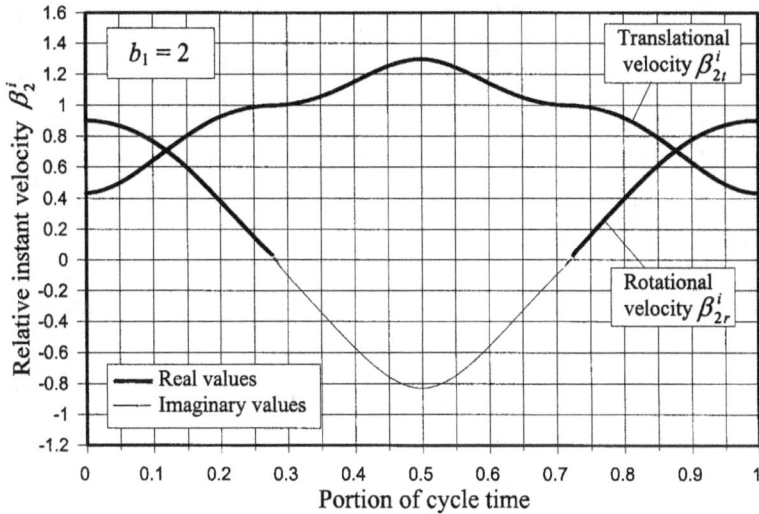

Figure 8.12.2. Variation of instant velocities β_{2t}^i and β_{2r}^i of the trailing string within one cycle of trailing string.

Figure 8.12.2 shows the variation of the relative instant translational and rotational velocities, β_{2t}^i and β_{2r}^i for the case when $b_1 = 2$. Notably, between 0.25 and 0.75 portions of the cycle of the trailing string, β_{2t}^i exceeds the velocity of light while β_{2r}^i is expressed with imaginary numbers.

8.13 Toryx Golden Polarization Factor - The toryx golden polarization factor G_t is equal to the product of the toryx vorticity V and a square root of the toryx reality R with an opposite sign.

From Eqs. (8.1-1) and (8.2-1) this factor is defined by the equation:

$$G_t = -V\sqrt{R} = \frac{(b_1 - 1)\sqrt{2b_1 - 1}}{b_1} \tag{8.13-1}$$

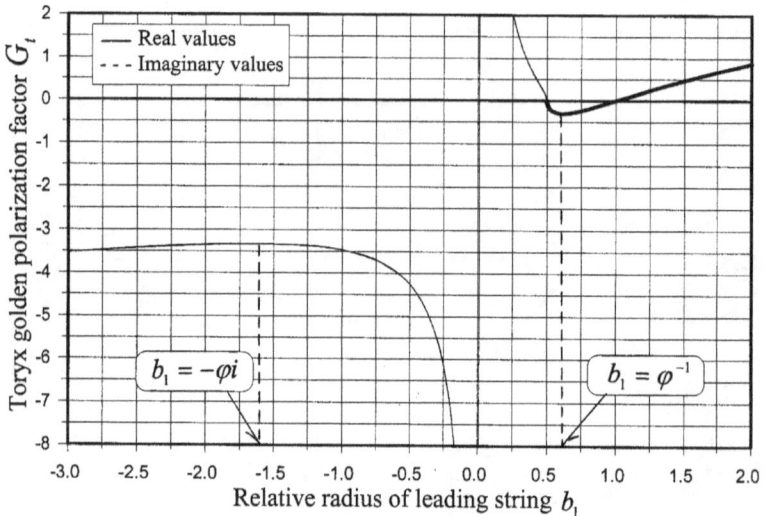

Figure 8.13. Toryx golden polarization factor G_t as a function of the relative radius of leading string b_1.

Figure 8.13 shows a plot of Eq. (8.13-1) in which the toryx golden polarization factor G_t is expressed as a function of the relative radius of leading string b_1.

Notably, the plot of G_t has two extreme values of G_t at which b_1 is directly related to the golden ratio φ. The first extreme value of G_t corresponds to $b_1 = -\varphi i$; here the value of G_t is maximum. The second extreme value of G_t corresponds to $b_1 = \varphi^{-1}$; here the value of G_t is minimum.

CHAPTER 9

INVERSION STATES OF TORYX

The branch of sciences that studies inversion of curved spaces is called *topology*. One of the most famous topologists of the 19th century was the German mathematician and astronomer Ferdinand Möbius (1790-1868). To illustrate the idea of inverted surface, Möbius described a topological shape that became known as the *Möbius strip* in a paper discovered only after his death. Take a piece of paper, rotate one end through 180 degrees and connect the ends together (Fig. 9).

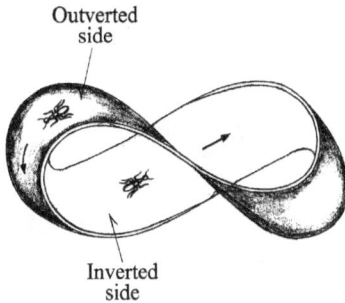

Outverted
side

Inverted
side

Figure 9.
An ant walking along
the Möbius strip.

An ant walking on the inside part of the Möbius strip will eventually end up on the outer part of the strip. Thus, the Möbius strip makes a smooth transition from the inverted surface to the surface turned inside out. For the ant, the Möbius strip is a two-dimensional surface. Let us now consider the inversions of the toryx leading string, trailing string and spherical membrane.

9.1 Inversion of Leading String - To illustrate the inversion of its circular leading string let us envision it in the form of an extremely thin and narrow circular ribbon with the relative radius b_1 (Fig. 9.1).

Let us assume that the leading string is propagating outward when b_1 decreases from positive infinity $(+\infty)$. In that case, the outer color of the ribbon is black, while its inner color is white. As the ribbon radius b_1 decreases while remaining positive, the ribbon begins to appear smaller, but still propagating outward until b_1 reduces to positive infinity $(+0)$. Then the leading string inverts making b_1 approaching negative infinity (-0). As the negative values of b_1 increase and extend to negative infinity $(-\infty)$, its outer color of the leading string appears white, while its inner color looks black.

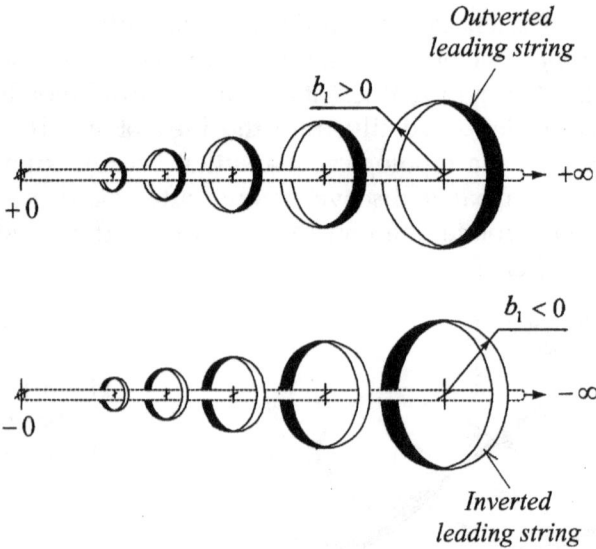

Fig. 9.1. Inversion of a leading string.

9.2 Inversion of Trailing String - To illustrate the inversion of the toryx trailing string let us envision it in the form of an extremely thin and narrow toroidal ribbon (Fig. 9.2).

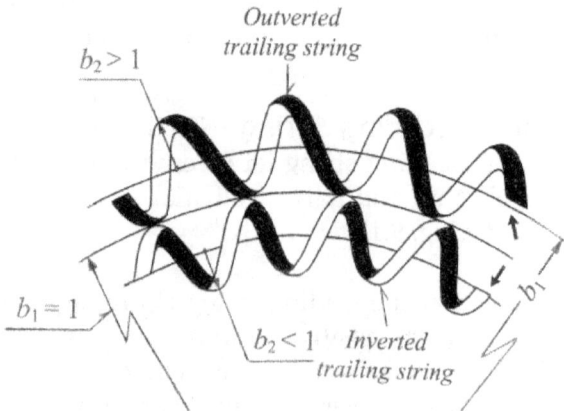

Fig. 9.2. Inversion of a trailing string.

We assume that trailing string propagates outward when the relative radius of leading string b_1 is greater than 1. In this case, the radius of the trailing string b_2 is positive and the outer color of the toroidal ribbon is black. When b_1 becomes infinitesimally close to 1, the radius of trailing string b_2 approaches positive infinility (+0). So, the toryx reduces to a circle seen in Figure 9.2 as a colorless edge of a circular ribbon the radius $b_1 = 1$. When b_1 becomes less than 1, the sign of the radius b_2 changes from positive to negative and the ribbon becomes inverted. Consequently, its outer color becomes white while its inner color becomes black.

9.3 Inversion of Spherical Membrane – To simplify the picture, envision the spherical membrane in the form of an extremely thin spherical membrane with the relative radius b (Fig. 9.3).

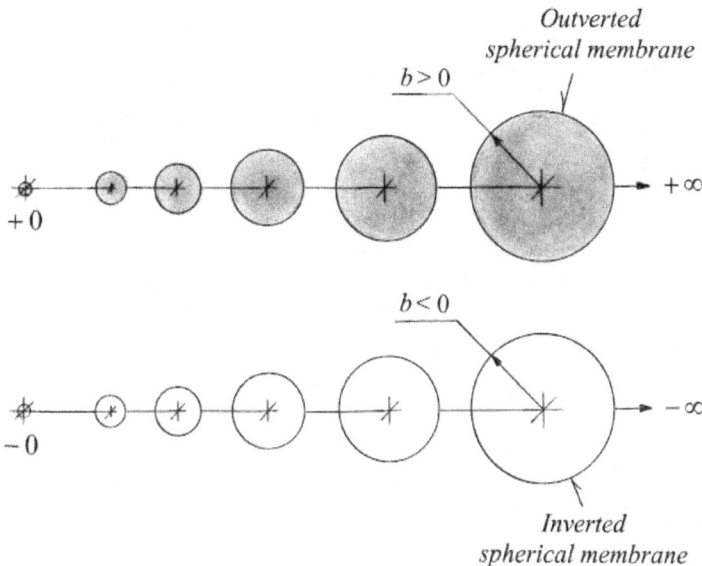

Fig. 9.3. Inversion of a spherical membrane.

Let us start from the outverted spherical membrane in which b decreases from positive infinity (+∞). In that case, its outer color is assumed to be black, while its inner color is white. As b decreases while remaining positive, the spherical membrane begins to appear smaller, and it eventually reduces, so b approaches positive infinility (+0). Then the spherical membrane inverts, so b approaches negative infinility (−0). As the negative values of b increase and extend to negative infinity (−∞), the outer color of the spherical membrane appears white, while its inner color looks black.

9.4 Inversion of Toryces - Figure 9.4 shows metamorphoses of the toryx leading string with the relative radius b_1 and trailing strings with the relative radius b_2 as the steepness angle of the toryx trailing string φ_2 increases from 0^0 to 360^0.

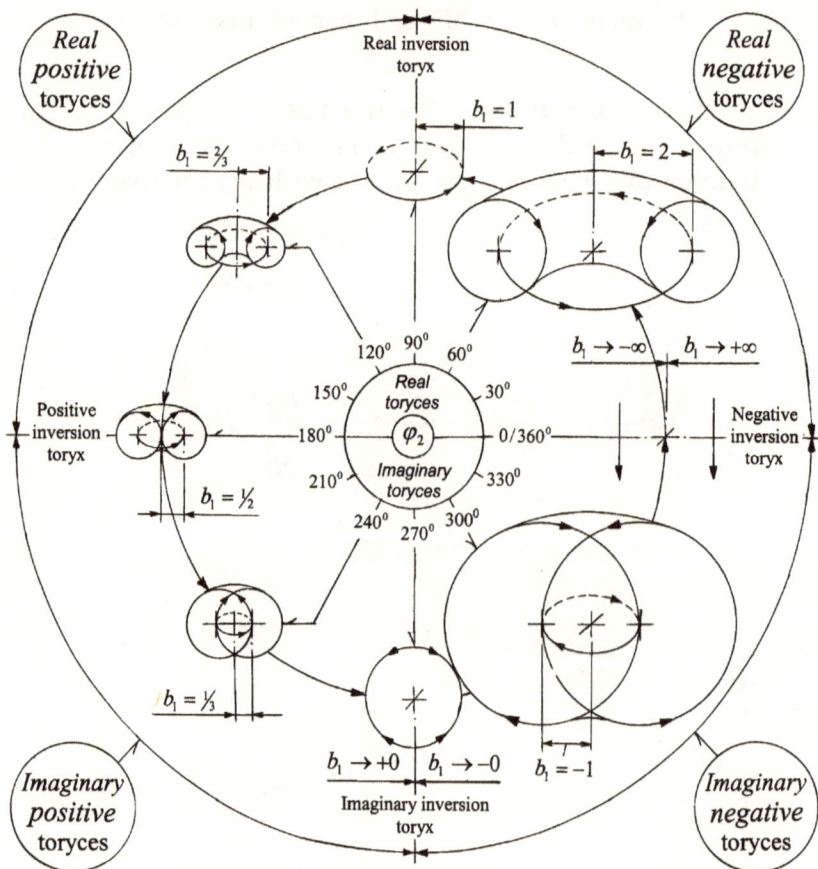

Figure 9.4. Metamorphoses of the toryx leading and trailing strings as a function of the steepness angle of trailing string φ_2.

Four kinds of *inversion toryces* are located at the boundaries of the circular diagram between the four main groups of toryces.

- Negative inversion toryx ($\varphi_2 \to +0^0 / 360^0$) - At this point, both leading and trailing strings become inverted. Consequently, $b_1 \to \pm\infty$ and $b_2 \to \pm\infty$. The toryx appears as two parallel lines separated by the distance equal to the diameter of real inversion string.

- Real inversion toryx ($\varphi_2 \to 90^0$) - At this point, only trailing string becomes inverted. Consequently, $b_1 \to +1$, $b_2 \to \pm0$ and the toryx appears as circle with the relative radius $b_1 \to +1$.

- Positive inversion toryx ($\varphi_2 \to 180^0$) - At this point, only spherical membrane becomes inverted. Consequently, $b_1 \to +\frac{1}{2}$, $b_2 \to -\frac{1}{2}$ and the toryx appears as an extreme case of a *spindle torus* with inner parts of its windings touching one another.

- Imaginary inversion toryx ($\varphi_2 \to 270^0$) - At this point, only the leading string becomes inverted. Consequently, $b_1 \to \pm0$, $b_2 \to -1$ and the toryx appears as a circle with the relative radius approaching -1. The circle is located at the plane perpendicular to the plane of the real inversion string

9.5 Transformations of Toryces Within Each Group - Located between the inversion toryces on the circular diagram of Figure 9.4 are the toryces that belong to their four main groups.

Figures 9.5.1 – 9.5.4 show the transformations of the toryces within each main group.

Real negative toryces (Fig. 9.5.1) - These toryces belong to the top right quadrant of the circular diagram shown in Figure 9.4. Trailing strings of these toryces are wound counter-clockwise outside of the real inversion string. As φ_2 increases, both b_1 and b_2 decrease, so that the trailing string appears like a conventional torus.

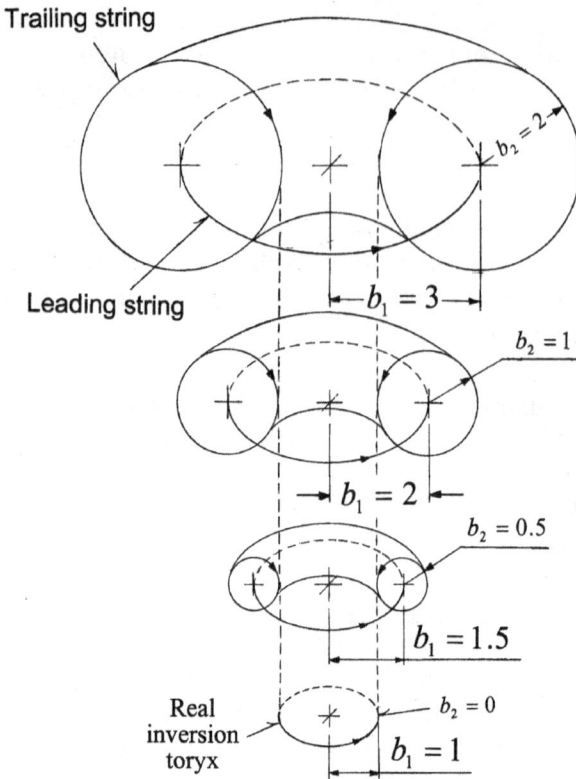

Figure caption content below. The figure contains labels: Trailing string, Leading string, $b_2 = 2$, $b_1 = 3$, $b_2 = 1$, $b_1 = 2$, $b_2 = 0.5$, $b_1 = 1.5$, $b_2 = 0$, $b_1 = 1$, Real inversion toryx.

Ranges of toryx spacetime parameters.

Range	φ_2	b	b_1	b_2	η_2	w_2	β_{2t}	β_{2r}
From	0^0	$+\infty$	$+\infty$	$+\infty$	$+\infty$	$+\infty$	$+0$	1.0
To	90^0	1.0	1.0	$+0$	1.0	1.0	1.0	$+0$

Figure 9.5.1. Transformation of the real negative toryx.

Real positive toryces (Fig. 9.5.2) - These toryces belong to the top left quadrant of the circular diagram shown in Figure 9.4. Within this range the trailing string is inverted, so that its windings are now wound clockwise inside the real inversion toryx. As φ_2 increases, b_1 decreases while the negative value of b_2 increases. Consequently, the toryx appears as an inverted toroidal spiral.

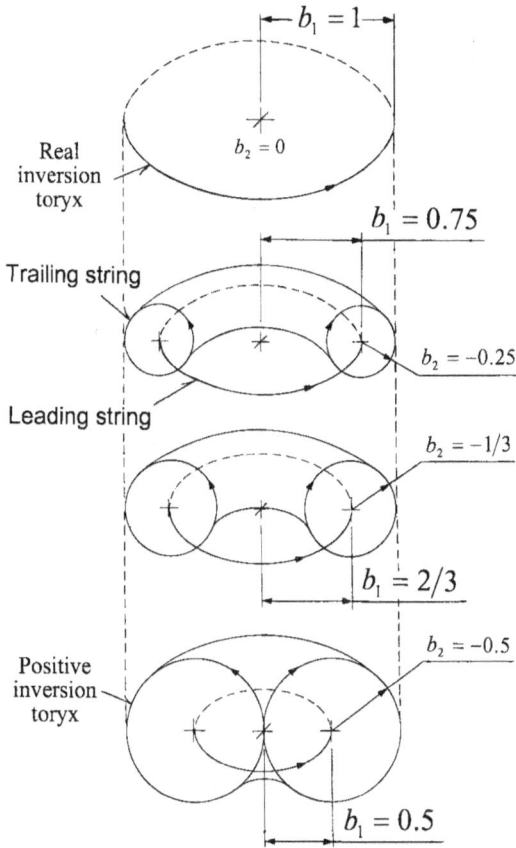

Ranges of toryx spacetime parameters.

Range	φ_2	b	b_1	b_2	η_2	w_2	β_{2t}	β_{2r}
From	90^0	1.0	1.0	-0	1.0	1.0	1.0	-0
To	180^0	$+0$	½	$-½$	$+0$	$+\infty$	$+0$	-1.0

Figure 9.5.2. Transformation of the real positive toryx.

Imaginary positive toryces (Fig. 9.5.3) - These toryces belong to the bottom left quadrant of the circular diagram shown in Figure 9.4. As φ_2 increases, b_1 decreases while negative values of b_2 increase. Within this range, the opposite parts of windings of the trailing string intersect with one another like in a *spindle torus*.

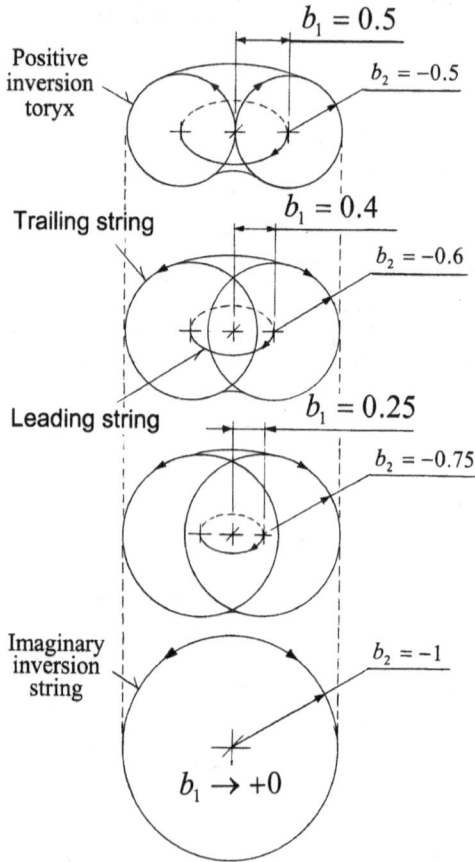

Positive inversion toryx

$b_1 = 0.5$

$b_2 = -0.5$

Trailing string

$b_1 = 0.4$

$b_2 = -0.6$

Leading string

$b_1 = 0.25$

$b_2 = -0.75$

Imaginary inversion string

$b_2 = -1$

$b_1 \rightarrow +0$

Ranges of toryx spacetime parameters.

Range	φ_2	b	b_1	b_2	η_2	w_2	β_{2t}	β_{2r}
From	180^0	-0	$\frac{1}{2}$	$-\frac{1}{2}$	$-0i$	$-\infty i$	$-0i$	-1.0
To	270^0	-1.0	$+0$	-1.0	$-i$	$-0i$	$-\infty i$	$-\infty$

Figure 9.5.3. Transformation of the imaginary positive toryx.

Imaginary negative toryces (Fig. 9.5.4) - These toryces belong to the bottom right quadrant of the circular diagram shown in Figure 9.4. Here the leading string becomes inverted. As the negative value of b_1 increases, the negative values of b_2 also increase. Within this range the toryx windings are located outside of the imaginary inversion toryx.

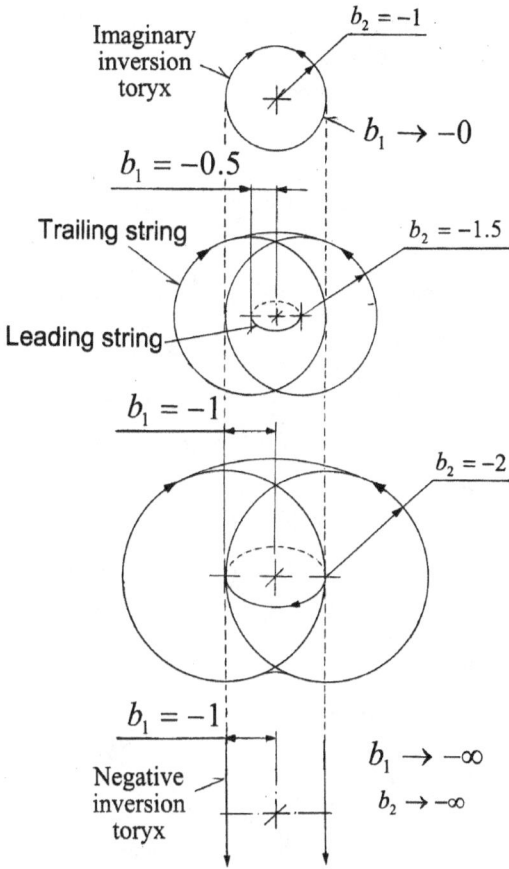

Ranges of toryx spacetime parameters.

Range	φ_2	b	b_1	b_2	η_2	w_2	β_{2t}	β_{2r}
From	270^0	-1.0	-0	-1.0	$-i$	$+0i$	$-\infty i$	$+\infty$
To	360^0	$-\infty$	$-\infty$	$-\infty$	$-\infty i$	$+\infty i$	$-0i$	1.0

Figure 9.5.4. Transformation of the imaginary negative toryces.

9.6 Summary of Toryx Transformations – Figure 9.6 provides a complete overview of transformation of toryces by showing together the metamorphoses of the toryx spherical membrane, leading string and trailing string with the respective relative radii b, b_1 and b_2.

Figure 9.6 shows the transformations of the toryx leading and trailing strings and its spherical membrane as the steepness angle of the trailing string φ_2 increases from 0^0 to 360^0.

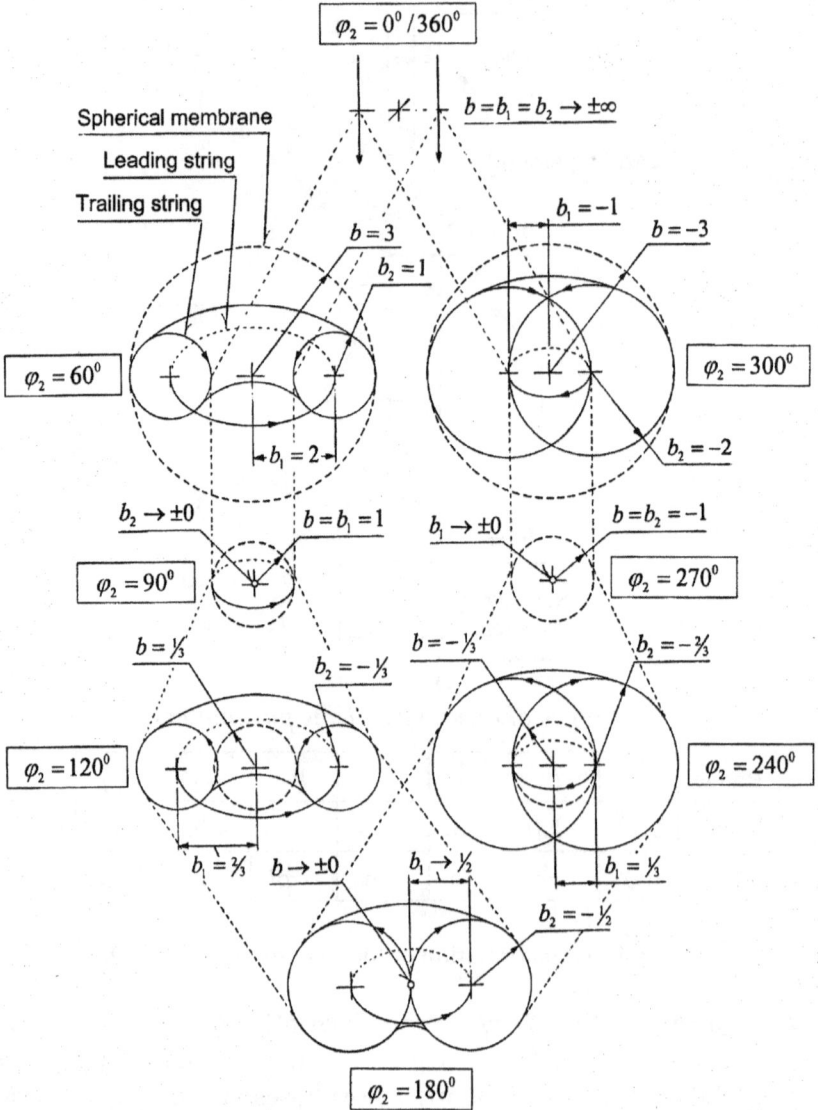

Figure 9.6. Transformations of the toryx leading and trailing strings and its spherical membrane.

- $\varphi_2 = +0^0$ – At this point, all three radii b_1, b and b_2 approach positive infinity $(+\infty)$.
- $+0^0 < \varphi_2 < 90^0$ - Within this range, b_1, b and b_2 decrease as φ_2 increases.
- $\varphi_2 \rightarrow 90^0$ – At this point, $b = b_1 = 1$ and b_2 approaches positive infinity $(+0)$; *trailing strings becomes inverted* and b_2 becomes negative.
- $90^0 < \varphi_2 < 180^0$ - Within this range, b_1 and b continue to decrease, while negative values of b_2 increase starting from negative infinity (-0).
- $\varphi_2 \rightarrow 180^0$ – At this point, $b_1 = \frac{1}{2}$, $b_2 = -\frac{1}{2}$ and b approaches positive infinity $(+0)$; *spherical membrane becomes inverted* and b becomes negative.
- $180^0 < \varphi_2 < 270^0$ - Within this range, b_1 continues to decrease, negative values of b_2 continue to increase, while negative values of b increase starting from negative infinity (-0).
- $\varphi_2 \rightarrow 270^0$ – At this point, $b = b_2 = -1$, while b_1 approaches positive infinity $(+0)$; *leading string becomes inverted* and b_1 becomes negative.
- $270^0 < \varphi_2 < 360^0$ - Within this range, negative values of b_1 increase starting from negative infinity (-0), while negative values of b and b_2 continue to increase.
- $\varphi_2 \rightarrow 360^0$ – At this point, all three radii b, b_1 and b_2 approach negative infinity $(-\infty)$.

The transformations of the toryx leading string and trailing strings and its spherical membrane are summarized in Table 9.6:

- The inversions of the toryx leading string, trailing string and spherical membrane occur when their radii approach either infinity (± 0) or infinity $(\pm \infty)$.
- Relative radius of the toryx trailing string b_2 approaches infinity (± 0) when $\varphi_2 \rightarrow 90^0$.
- Relative radius of the toryx spherical membrane b approaches infinity (± 0) when $\varphi_2 \rightarrow 180^0$.
- Relative radius of the toryx leading string b_1 approaches infinity (± 0) when $\varphi_2 \rightarrow 270^0$.

- Relative radii of spherical membrane b, leading string b_1 and trailing string b_2 approach infinity $(\pm\infty)$ when $\varphi_2 \to 0^0/360^0$.
- Relative radii of the toryx leading string b_1 and spherical membrane b approach positive unity $(+1)$ when $\varphi_2 \to 90^0$.
- Relative radii of the toryx trailing string b_2 and spherical membrane b approach negative unity (-1) when $\varphi_2 \to 270^0$.

Table 9.6. Inversion points of the toryx leading string and trailing strings and its spherical membrane.

Toryx component	Relative radius	Steepness angle
Spherical membrane	$b \to \pm\infty$	$\varphi_2 \to 0^0/360^0$
	$b \to \pm 0$	$\varphi_2 \to 180^0$
Leading string	$b_1 \to \pm\infty$	$\varphi_2 \to 0^0/360^0$
	$b_1 \to \pm 0$	$\varphi_2 \to 270^0$
Trailing string	$b_2 \to \pm\infty$	$\varphi_2 \to 0^0/360^0$
	$b_2 \to \pm 0$	$\varphi_2 \to 90^0$

CHAPTER 10

PHYSICAL PROPERTIES OF MICRO-TORYCES

Our theory is based on the assumption that both space and time are the universal terms. The history of humanity tells us that since the early times of existence of human civilization people quickly learned how to measure the distances between various objects by comparing these distances, for instance, with the lengths of their feet. Thanks to the periodicity of sunsets they were also able to measure a passage of time by counting the number of days. The same discoveries could be made by intelligent beings that might exist in any part of the universe.

It would, however, be a completely different story when it comes to describing physical properties of matter. Physical terms, such as mass, charge, force, energy, magnetic induction, etc. are purely man-made inventions based on their theories. It is very likely, however, that the other intelligent beings developed completely different physical theories of nature and, consequently, use completely different physical terms. Our theory expresses physical properties of the micro-toryces as a function of spacetime parameters that can be comprehendible by any intelligence in the universe.

Another important feature of our theory is that the toryces are applicable to the entities of both micro- and macro-worlds. They are respectively called *micro-toryces* and *macro-toryces*. There is one principal difference between the micro- and macro-toryces. The toryces of the micro-world describe the properties of prime elements of nature making up elementary particles, while the toryces of the macro-world describe the properties of fields associated with the entities of the macro-world. The spacetime properties of both micro- and macro-toryces are described by the same spacetime equations, except for the equation describing the radii of their real inversion toryces. To make a distinction between these radii, we use the symbols r_i for micro-toryces and the symbol r_j for macro-toryces.

10.1 Radius of Real Inversion Toryx – The radius of the real inversion toryx r_i is one of the major parameters allowing us to establish a correlation between physical and spacetime properties of the micro-toryces.

The real inversion toryx is shown at the top of Figure 9.4. It corresponds to the case when the relative radius of leading string $b_1 = 1$. Thus, in application to the micro-toryx the radius of real inversion toryx r_i is equal to the toryx eye radius r_0.

$$r_i = r_0 \qquad (10.1\text{-}1)$$

It possible to express the radius r_i in physical terms by comparing the velocity of the toryx spiral string V_1 based on our theory with the orbital velocity of atomic electron based on the classical law of planetary motion. We find from Eq. (5.7-8) that the relative spiral velocity β_1 of the toryx leading string is equal to:

$$\beta_1 = \frac{V_1}{c} = \frac{\sqrt{2b_1 - 1}}{b_1} \qquad (10.1\text{-}2)$$

Classical law of planetary motion for atomic electron can be derived by equating two forces:

- The attraction electric force F_e between the positive electric charge Ze, where Z is the atomic number, and the negative electric charge e separated by the distance r_1

- The centrifugal forces F_c applied to the electron with the mass m_e orbiting the nucleon with the orbital velocity V_1 (Fig. 10.1.1).

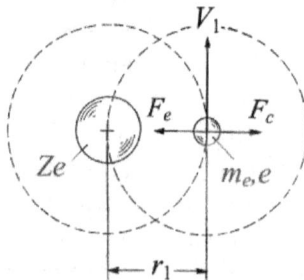

Figure 10.1.1. Atomic planetary system with two charged particles.

The equilibrium between the electric force F_e and the centrifugal forces F_c will exist when the two conditions are met. Firstly, the toryx eye radius r_0 must be equal to the radius of the real inversion toryx r_i:

$$r_0 = r_i = \frac{Ze^2}{8\pi\varepsilon_0 m_e c^2} \qquad (10.1\text{-}3)$$

Secondly, the relative radius of the toryx leading string must be equal to:

$$\beta_1 = \frac{V_1}{c} = \sqrt{\frac{2}{b_1}} \qquad (10.1\text{-}4)$$

The above equation describes the law of planetary motion of classical mechanics in which both the velocity and distance are expressed in relative terms. One may readily see that this equation is a particular case of Eq. (10.1-2) applied to the toryx leading string when $b_1 \gg 1$. Therefore, the classical law of planetary motion for atomic electron is merely a particular case of a more general law of planetary motion applied to the toryx that we call the *Unified Law of planetary Motion.*

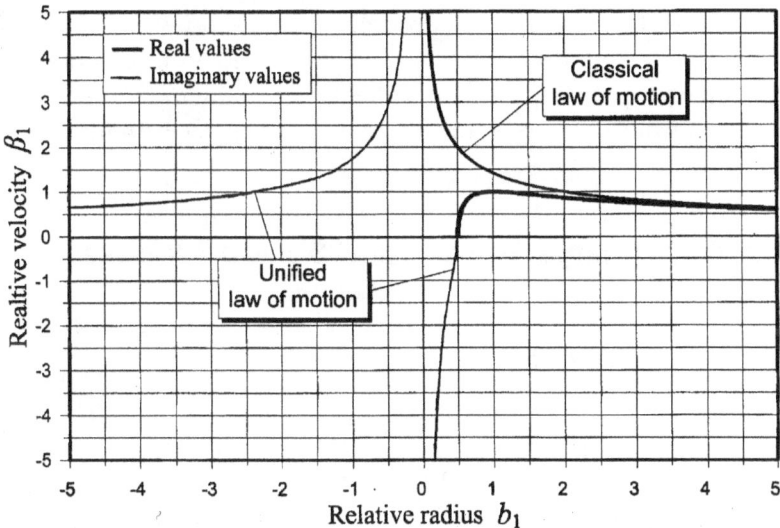

Figure 10.1.2. Unified law of planetary motion versus classical law of planetary motion.

Figure 10.1.2 shows the plots of Eqs. (10.1-2) and (10.1-4) representing respectively unified and classical laws of planetary motion. The highlights of the plot are:

- Classical law of planetary motion is applied to the range of b_1 extending from zero to positive infinity; all values of β_1 are expressed with real numbers.
- Unified law of planetary motion is applied to the range of b_1 extending from negative to positive infinity; within the range of b_1 extending from 0.5 to positive infinity the values of β_1 are expressed with real numbers, while within the remaining range of b_1 with imaginary numbers.
- When $b_1 > 5$, the difference between the calculated values of β_1 from the unified and classical laws of planetary motion becomes small and it continues to decrease as b_1 increases.
- As b_1 decreases from 5 to 2, this difference progressively increases.
- As b_1 decreases from 2 to 0, according to the classical law of planetary motion, β_1 sharply increases within this range and approaches positive infinity $(+\infty)$.
- As b_1 decreases from 2 to 0, according to the unified law of planetary motion, β_1 initially increases within the same range and then, after reaching its maximum value of 1 (corresponding to the velocity of light c) at $b_1 = 1$, it sharply decreases and approaches negative imaginary infinility $(+0i)$.

10.2 Toryx Charge, Gravitational & Inertial Masses

Toryx charge - The toryx relative charge e_t/e is assumed to be equal to the toryx vorticity V with opposite sign.

$$\frac{e_t}{e} = -V = -\frac{b_1 - 1}{b_1} = -\frac{b - 1}{b + 1} = -\cos u\varphi_2 \qquad (10.2\text{-}1)$$

Toryx gravitational mass - The toryx relative gravitational mass m_{tg}/m_e is assumed to be equal to the absolute value of the toryx vorticity V.

$$\frac{m_{tg}}{m_e} = |V| = \left|\frac{b_1 - 1}{b_1}\right| = \left|\frac{b - 1}{b + 1}\right| = |\cos u\varphi_2| \qquad (10.2\text{-}2)$$

Toryx inertial mass – The toryx relative inertial mass m_{ti}/m_e is assumed to be equal to the toryx vorticity V.

$$\frac{m_{ti}}{m_e} = \frac{b_1 - 1}{b_1} = \frac{b-1}{b+1} = \cos u\varphi_2 \qquad (10.2\text{-}3)$$

The UST yields a completely different set of relativistic equations than those used in the Albert Einstein's special theory of relativity. The main features of the toryx relativistic equations shown in Table 10.2 are:

- Not only masses of the toryces but also their charges are dependent on velocities of their leading strings.
- In the real toryces the absolute values of their relative charges and masses decrease with an increase of relative spiral velocities of their leading strings β_1.
- Conversely, in the imaginary toryces the absolute values of their relative charges and masses increase as the relative spiral velocities of their leading string β_1.

Table 10.2 Relativistic equations for the toryx relative charge and gravitational mass.

Toryx parameters	Equations
Relative charge	$\dfrac{e_t}{e} = \sqrt{1 - \beta_1^2}$ (10.2-4)
Relative gravitational mass	$\dfrac{m_{tg}}{m_e} = \left\| \sqrt{1 - \beta_1^2} \right\|$ (10.2-5)

The frequency of the real inversion toryx f_i and period T_i are equal to:

$$f_i = f_0 = \frac{4\varepsilon_0 m_e c^3}{Ze^2} \qquad (10.2\text{-}6)$$

$$T_i = T_0 = \frac{Ze^2}{4\varepsilon_0 m_e c^3} \qquad (10.2\text{-}7)$$

10.3 Toryx Mechanical Properties - Described below are three toryx mechanical properties: density, elasticity and angular momentum. We defined these properties in a similar way as they would be defined by classical mechanics.

Toryx density – The toryx density ρ_t is equal to the ratio of the toryx gravitational mass m_{tg} to the volume occupied by its trailing string U_2, thus:

$$\rho_t = \frac{m_{tg}}{U_2} \qquad (10.3\text{-}1)$$

From Eqs. (5.7-11), (10.2-2) and (10.3-1), the toryx density ρ_t is equal to:

$$\rho_t = \frac{m_e}{2\pi^2 r_i^3} \frac{|b_1 - 1|}{b_1} \frac{1}{|b_1(b_1 - 1)^2} \qquad (10.3\text{-}2)$$

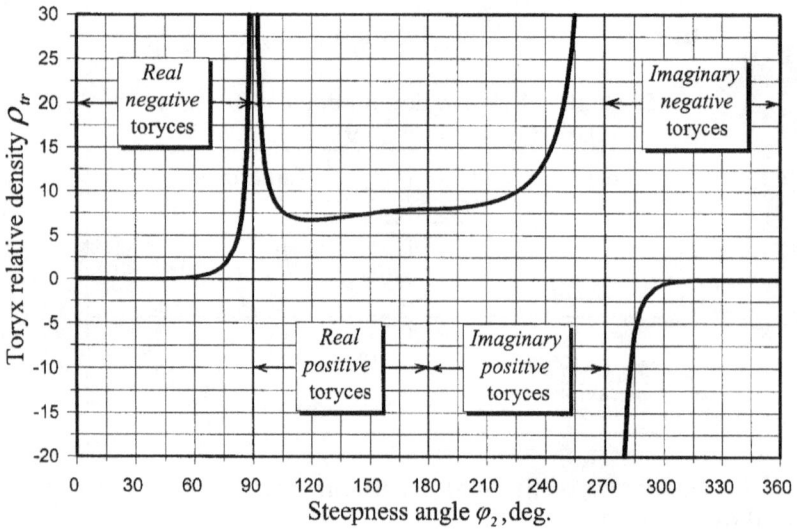

φ_2	$360^0/0^0$	90^0	180^0	270^0
ρ_{tr}	$-0/+0$	$+\infty$	8.0	$+\infty/-\infty$

Figure 10.3.1. Toryx relative density ρ_{tr} as a function of the steepness angle of trailing string φ_2.

From Eq. (10.3-2), the toryx relative density ρ_{tr} is equal to (see Fig. 10.3.1):

$$\rho_{tr} = \rho_t \frac{2\pi^2 r_i^3}{m_e} = \left| \frac{b_1 - 1}{b_1} \right| \frac{1}{b_1 (b_1 - 1)^2} \qquad (10.3\text{-}3)$$

Notably, the toryx density approaches the infinility (± 0) when φ_2 approaches 0^0 and 360^0, and it approaches the infinity ($\pm\infty$) when φ_2 approaches 90^0 and 270^0.

Toryx elasticity – In classical mechanics, elastic properties of compressible media contained in a certain volume are defined by the *bulk modulus of elasticity*. It is equal to the ratio of the change in pressure inside the media to the resulting fractional change in the volume of the media. Consider the toryx trailing string as a compressible media. Let compression waves, like sound waves, travel through the toryx trailing string with the velocity equal to the spiral velocity of the toryx leading string V_1. In that case, the toryx bulk modulus B_t will be equal to:

$$B_t = \rho_t V_1^2 \qquad (10.3\text{-}4)$$

From Eqs. (5.7-8), (10.3-1) and (10.3-4), the toryx bulk modulus of elasticity B_t is equal to:

$$B_t = \frac{m_e c^2}{2\pi^2 r_i^3} \left| \frac{b_1 - 1}{b_1} \right| \frac{2b_1 - 1}{b_1^3 (b_1 - 1)^2} \qquad (10.3\text{-}5)$$

The toryx relative bulk modulus of elasticity B_{tr} as a function of the steepness angle of trailing string φ_2 is equal to (see Fig. 10.3.2):

$$B_{tr} = B_t \frac{2\pi^2 r_i^3}{m_e c^2} = \left| \frac{b_1 - 1}{b_1} \right| \frac{2b_1 - 1}{b_1^3 (b_1 - 1)^2} \qquad (10.3\text{-}6)$$

Notably, the toryx relative bulk modulus of elasticity B_{tr} approaches infinility (± 0) when φ_2 approaches 0^0 and 360^0, and it approaches infinity ($\pm\infty$) when φ_2 approaches 90^0 and 270^0.

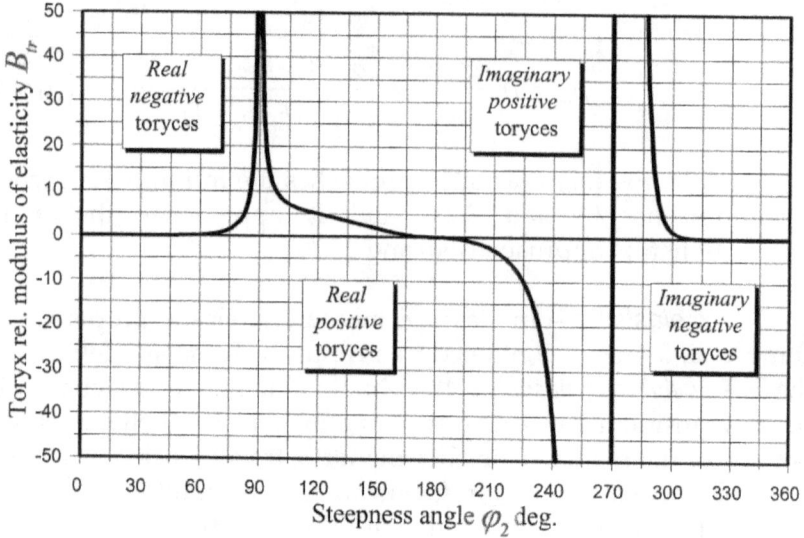

$\varphi 2$	$360^0/0^0$	90^0	180^0	270^0
B_{tr}	0	$+\infty$	$-0/+0$	$-\infty/+\infty$

Figure 10.3.2 Toryx relative bulk modulus of elasticity B_{tr} as a function of the steepness angle of trailing string φ_2.

Toryx angular momentum – Toryx angular momentum P_{ta} is defined by a typical classical equation:

$$P_{ta} = 2\pi r_1 m_{ti} V_1 \tag{10.3-7}$$

From Eqs. (5.7-1), (5.7-8), (10.2-3) and (10.3-7), the relative values of the toryx angular momentum are equal to (see 10.3.3):

$$\text{Real toryces: } P_{ta} = P_{ta} \frac{2\varepsilon_0 c^2}{Ze^2} = \pm \frac{(b_1-1)}{\sqrt{2b_1-1}_1} \tag{10.3-8}$$

$$\text{Imaginary toryces: } \breve{P}_{ta} = \breve{P}_{ta} \frac{2\varepsilon_0 c^2}{Ze^2} = \pm \frac{(b_1-1)}{i\sqrt{2b_1-1}_1} \tag{10.3-9}$$

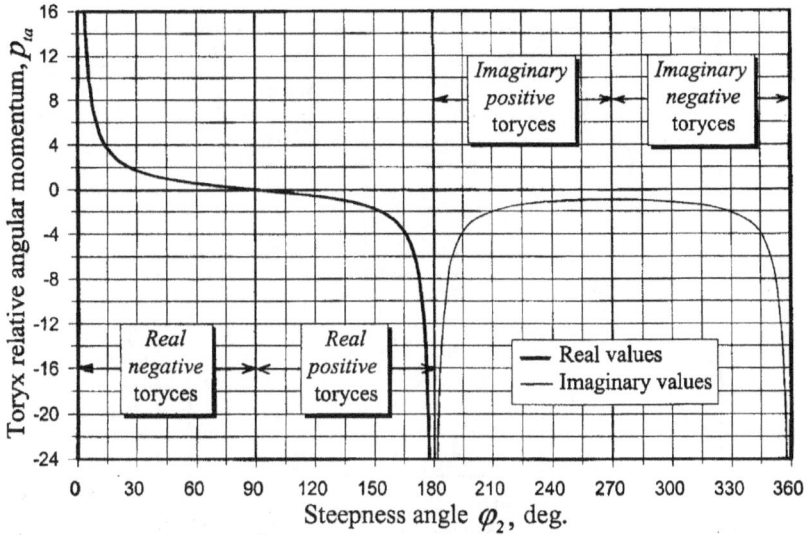

φ_2	$360^0/0^0$	90^0	180^0	270^0
P_{ta}	$-\infty i/+\infty$	$+0/-0$	$-\infty/-\infty i$	-1

Figure 10.3.3. Toryx relative angular momentum p_{ta} as a function of the steepness angle of trailing string φ_2 ($Z = 1$).

10.4 Toryx Energies – We defined below several kinds of toryx energies: kinetic, potential, exchange, field and matter.

Toryx kinetic energy - According to the Bohr's quantum model of the atom, the kinetic energy K of atomic electron is equal to:

$$K = \frac{m_{ti}V_1^2}{2}$$

(10.4-1)

From Eqs. (5.7-8), (10.2-3) and (10.4-1), the toryx kinetic energy K_t is equal to:

$$K_t = m_e c^2 \frac{(b_1-1)(2b_1-1)}{2b_1^3}$$

(10.4-2)

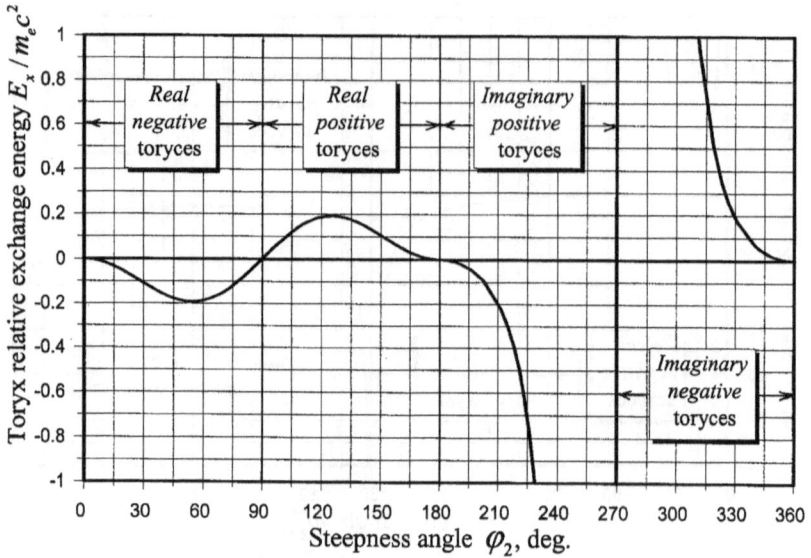

φ_2	$360^0/0^0$	90^0	180^0	270^0
E_x/m_ec^2	$+0/-0$	$-0/+0$	$+0/-0$	$-\infty/+\infty$

Figure 10.4.1. Toryx relative exchange energy E_x/m_ec^2 as a function of the steepness angle of trailing string φ_2.

Toryx potential energy - According to the Bohr's quantum model of the atom, the potential energy of atomic electron is equal to:

$$U = -2K \qquad (10.4\text{-}3)$$

Similarly, from Eqs. (10.4-2) and (10.4-3), the toryx potential energy U_t is equal to:

$$U_t = -m_ec^2\frac{(b_1-1)(2b_1-1)}{b_1^3} \qquad (10.4\text{-}4)$$

Toryx exchange energy - The toryx total kinetic and potential energy is called the toryx exchange energy E_x. From Eqs. (10.4-2) and (10.4-4) it is equal to (Fig. 10.4.1):

$$E_x = K_t + U_t = -m_ec^2\frac{(b_1-1)(2b_1-1)}{2b_1^3} \qquad (10.4\text{-}5)$$

Toryx field & matter energy – The toryx field & matter energy E_{fm} is equal to the sum of the toryx field energy E_f and the toryx matter energy E_m as defined by the equation:

$$E_{fm} = E_f + E_m = m_e c^2 \qquad (10.4\text{-}6)$$

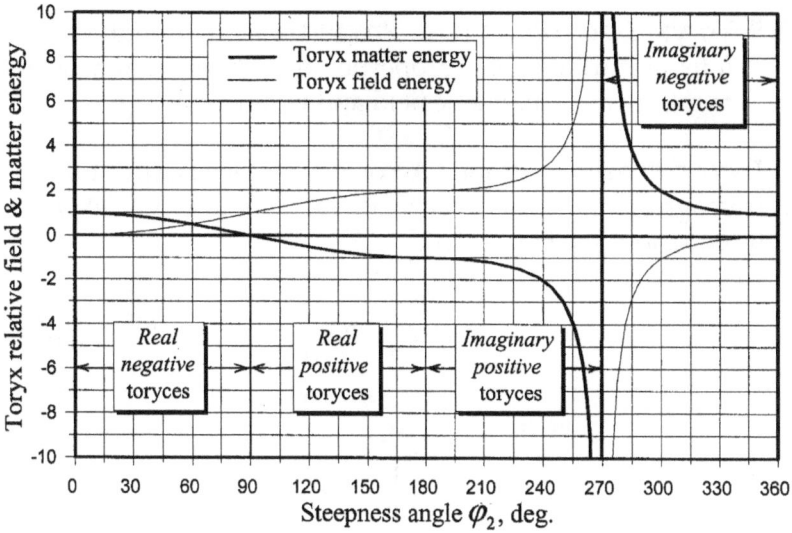

φ_2	$360^0 / 0^0$	90^0	180^0	270^0
$E_f/m_e c^2$	$-0/+0$	$+1$	$+2$	$+\infty/-\infty$
$E_m/m_e c^2$	$+1.0$	$+0/-0$	-1.0	$-\infty/+\infty$

Figure 10.4.2. Toryx relative field energy $E_f/m_e c^2$ and matter energy $E_m/m_e c^2$ as a function of the steepness angle of trailing string φ_2.

Toryx matter energy – The toryx matter energy E_m is associated with the toryx matter. It is defined by the equation:

$$E_m = m_e c^2 \frac{b_1 - 1}{b_1} \qquad (10.4\text{-}7)$$

Toryx field energy – The toryx field energy is associated with the toryx field. From equations (10.4-6) and (10.4-7), the toryx field energy E_f is equal to:

$$E_f = \frac{m_e c^2}{b_1} \tag{10.4-8}$$

Figure 10.4.2 shows plots of equations (10.4-7) and (10.4-8) as a function of the steepness angle of trailing string φ_2.

10.5 Toryx Electromagnetic Properties - We defined below four electromagnetic properties of toryces: magnetic moment, current, magnetic field and magnetic energy.

Toryx magnetic moment – According to the classical theory of electromagnetism the toryx magnetic moment μ_t can be expressed by the equation (see Fig. 10.5.1):

$$\mu_t = \frac{1}{2} e_t V_1 r_1 \tag{10.5-1}$$

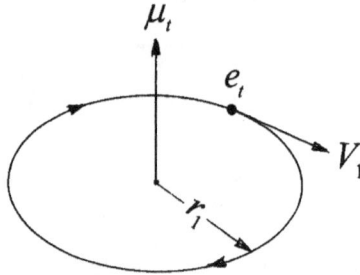

Figure 10.5.1. Toryx magnetic moment μ_t.

We express the toryx magnetic moment in relative terms in respect to either Bohr magneton μ_B or nuclear magneton μ_N. The Bohr magneton μ_B is given by the equation:

$$\mu_B = \frac{e^3}{8\pi\alpha\varepsilon_0 m_e c} \tag{10.5-2}$$

From Eqs. (5.7-1), (5.7-8), (10.2-1), (10.5-1) and (10.5-2) the toryx relative magnetic moment in respect to the Bohr magneton μ_t/μ_B:

Real toryces: $\quad \dfrac{\mu_t}{\mu_B} = \pm\alpha \dfrac{(b_1-1)\sqrt{2b_1-1}}{2b_1}$ $\qquad$ (10.5-3)

Imaginary toryces: $\quad \dfrac{\breve{\mu}_t}{\mu_B} = \pm\alpha \dfrac{(\breve{b}_1-1)i\sqrt{2\breve{b}_1-1}}{2\breve{b}_1}$ $\qquad$ (10.5-4)

The toryx relative magnetic moment in respect to the nuclear magneton μ_t/μ_N is related to the toryx relative magnetic moment in respect to the Bohr magneton μ_t/μ_B by the equation:

$$\frac{\mu_t}{\mu_N} = \frac{\mu_t}{\mu_B}\frac{m_p}{m_e}$$ (10.5-5)

where m_p is proton rest mass.

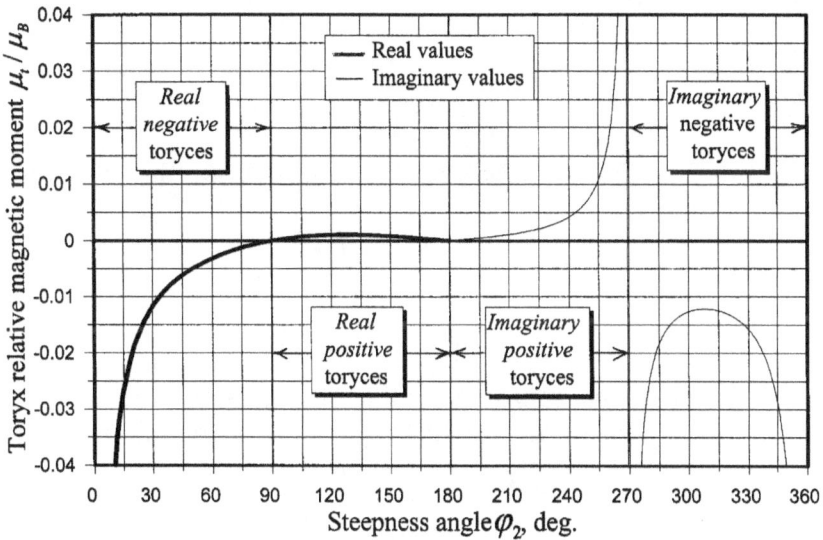

φ_2	$360^0/0^0$	90^0	180^0	270^0
μ_t/μ_B	$-\infty i/+\infty$	$+0/-0$	$-0/+0i$	$+\infty i/-\infty i$

Figure 10.5.2. Toryx relative magnetic moment in respect to the Bohr magneton μ_t/μ_B as a function of the steepness angle of trailing string φ_2.

Figure 10.5.2 shows a plot of equations (10.5-3) and (10.5-4) as a function of the steepness angle of trailing string φ_2. The relative magnetic moment μ_t/μ_B reaches its minimum value when the relative radius of leading string b_1 is equal to the inverse of the golden ratio.

From Eqs. (8.13-1), (10.5-3) and (10.5-4), the toryx relative magnetic moment in respect to the Bohr magneton μ_t/μ_B is proportional to the toryx golden polarization factor G_t.

Real toryces:
$$\frac{\mu_t}{\mu_B} = \pm \frac{\alpha G_t}{2} \qquad (10.5\text{-}6)$$

Imaginary toryces:
$$\frac{\breve{\mu}_t}{\mu_B} = \pm \frac{\alpha G_t i}{2} \qquad (10.5\text{-}7)$$

Toryx current – According to the classical theory of electricity, the toryx current I_t can be expressed by the equation:

$$I_t = \frac{e_t V_1}{2\pi r_1} \qquad (10.5\text{-}8)$$

From Eqs. (5.7-1), (5.7-8), (10.2-1) and (10.5-8), the toryx current I_t is equal to:

$$I_t = -\frac{ec}{2\pi r_i} \frac{(b_1 - 1)\sqrt{2b_1 - 1}}{b_1^3} \qquad (10.5\text{-}9)$$

Toryx magnetic field – According to the Ampère's law, the toryx magnetic field B_{tm} created by the toryx current I_t at the distance r_2 from the current line is given by the equation:

$$B_{tm} = \frac{\mu_0 I_t}{2\pi r_2} \qquad (10.5\text{-}10)$$

From Eqs. (5.7-1), (10.5-9) and (10.5-10), the toryx magnetic field B_{tm} is equal to:

$$B_{tm} = -\frac{\mu_0 ec}{4(\pi r_i)^2} \frac{\sqrt{2b_1 - 1}}{b_1^3} \qquad (10.5\text{-}11)$$

Toryx magnetic energy – The toryx magnetic energy E_{tm} corresponding to the toryx magnetic field B_{tm}, the volume occupied by the trailing string U_t and the radius of trailing string r_2 can be expressed by the equation:

$$E_{tm} = \frac{B_{tm}^2 U_t}{2\pi r_2}$$ (10.5-12)

From Eqs. (5.7-1),(10.5-11) and (10.5-12), the toryx magnetic energy E_{tm} is equal to:

$$E_{tm} = \frac{\pi^2 r_i^3 B_{tm}^2}{\mu_0} b_1 (b_1 - 1)^2$$ (10.5-13)

CHAPTER 11

FORMATION OF MATTER PARTICLES

Structure of elementary particles and their classification was a principle subject of the quark model developed during the second half of the 20th century as briefly described in Chapter 1. The UST describes this subject differently.

11.1 Classification of Particles – Our theory divides particles into two groups: *elementary* and *composite*. The elementary particles are called *trons*. Each tron is made up of two polarized toryces, while each composite particle is composed from two or more trons. Neutron and proton are the examples of composite particles.

According to the UST, there are only four elementary particles called *electrons, positrons, ethertrons* and *singulatrons*. Figure 11.1 shows a general presentation of formation of four elementary particles from polarized toryces.

- The ethertrons $a^0_{m,n,q}$ are charge-polarized real trons; they are made up of the real negative toryces $A^-_{m,n,q}$ and the real positive toryces $A^+_{m,n,q}$.
- The singulatrons $\breve{a}^0_{m,n,q}$ are charge-polarized imaginary trons; are made up of the imaginary negative toryces $\breve{A}^-_{m,n,q}$ and the imaginary positive toryces $\breve{A}^+_{m,n,q}$.
- The electrons $e^-_{m,n,q}$ are reality-polarized negative trons; they are made up of the real negative toryces $E^-_{m,n,q}$ and the imaginary negative toryces $\breve{E}^-_{m,n,q}$.
- The positrons $e^+_{m,n,q}$ are reality-polarized positive trons; they are made up of the real positive toryces $E^+_{m,n,q}$ and the imaginary positive toryces $\breve{E}^+_{m,n,q}$.

We use the same letters for symbols of trons and their constituent toryces; the lower-case letters are used for trons and capital letters for toryces. The superscripts in the symbols of toryces and trons indicate the signs of their charges and, in some cases, the magnitudes of these charges. The subscripts indicate the toryx quantum energy states m, n and q. We will explain physical meaning of these states later.

Ethertron

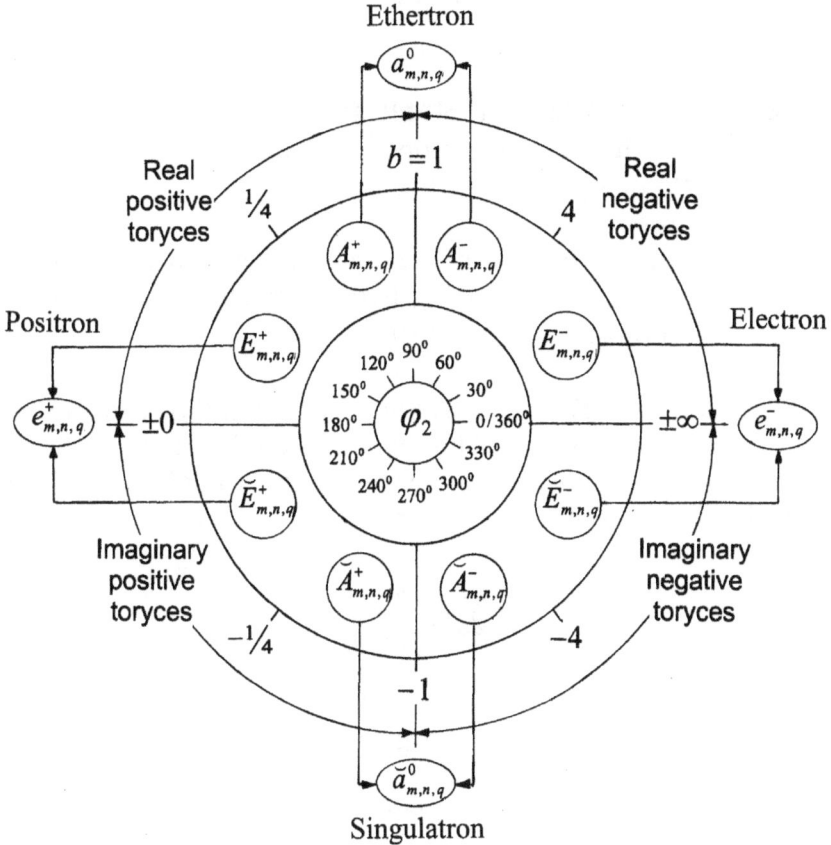

Figure 11.1. General presentation of formation of four elementary matter particles (trons) from polarized toryces.

For an elementary particle with the relative charge ε the relative radii of leading strings its constituent toryces b_1' and b_1'' are related to one another by the equation:

$$b_1'' = \frac{b_1'}{2b_1'(\varepsilon + 1) - 1} \tag{11.1-1}$$

The relative radii of spherical membranes of its constituent toryces b' and b'' are related to one another by the equation:

$$b'' = \frac{b'+1}{(b'+1)(\varepsilon+1)-1} - 1 \qquad (11.1\text{-}2)$$

11.2 The Unified Law of Stable Polarization – Particles are stable if they obey the proposed Unified Law of Stable Polarization.

According to this law, an enclosed system comprising of N toryces is stable if the following two polarization parameters of this system approach infinility (±0) :

Reality polarization – The reality polarization P_R is equal to the sum of the products of realities R_i and the relative gravitational masses m_{tg}/m_e of N toryces, and it must approach infinility (±0) as given by the equation:

$$P_R = \sum_{i=1}^{N} R_i \left(\frac{m_{tg}}{m_e} \right)_i \rightarrow \pm0 \qquad (11.2\text{-}1)$$

Charge polarization - The charge polarization P_C is equal to the sum of the relative charges e_t/e of N toryces, and it must approach infinility (±0) as given by the equation:

$$P_C = \sum_{i=1}^{N} \left(\frac{e_t}{e} \right)_i \rightarrow \pm0 \qquad (11.2\text{-}2)$$

Here is how the Unified Law of Stable Polarization affects the relationships between the relative radii of leading strings of the tron's constituent toryces.

In the reality-polarized trons, the relative radii of spherical membranes of their constituent toryces are related to one another by the equations:

$$\breve{b}^- = -b^- \qquad (11.2\text{-}3)$$

$$\breve{b}^+ = -b^+ \qquad (11.2\text{-}4)$$

In the charge-polarized trons, the relative spheryx radii of their constituent real and imaginary toryces are related to one another by the equations:

$$b^+ = \frac{1}{b^-}$$

(11.2-5)

$$\tilde{b}^+ = \frac{1}{\tilde{b}^-}$$

(11.2-6)

Trons which constituent toryces obey Eqs. (11.2-3) – (11.2-6) are called *basic trons* and their constituent toryces are called *basic toryces.*

In addition, to follow the Unified Law of Stable Polarization, the number of real toryces with the relative radii of their spheryx b and leading string b_1 must be T times greater than the number of imaginary toryces. Thus, the *tron reality ratio T* is equal to:

$$T = \left(\frac{b+1}{b-1}\right)^2 = \left(\frac{b_1}{b_1-1}\right)^2$$

(11.2-7)

Consequently, the exchange energies of the imaginary toryces $\tilde{E}_x$ and the real toryces E_x are related to one another by the equation:

$$\tilde{E}_x = E_x T^4$$

(11.2-8)

To obey the Unified Law of Stable Polarization, the two polarized constituent toryces of a tron absorb and release exchange energy intermittently, reducing the total exchange energy of both toryces to infinity. Consequently, the magnitudes of physical parameters of the tron are equal to the arithmetic averages of the magnitudes of related parameters of its constituent toryces.

11.3 Excitation of Toryces – Similarly to the conventional electrons, the toryces change their dimensions in quantum steps.

During excitation of a toryx, the radius of its leading strings r_1 changes in quantum steps, while its eye radius r_0 remains constant and equal to the radius of the real inversion toryx r_i as shown in Figure 6.1.1.

Trons forming nucleons and atom are made up of the toryces with very specific excitation quantum states. These states are derived to meet a number of requirements:

- Charge, mass and magnetic moments of calculated nucleons and hydrogen atom must be comparable with the experimental values.
- The known stable atoms and their isotopes must follow the Unified Law of Stable Polarization.
- Neutron must decay into a free proton, an electron and an electron antineutrino.
- Unstable baryons must decay into either neutron or proton.
- When an electron of hydrogen atom is transferred from one quantum state to another, it must either emit or absorb photons with the frequencies defined by the Rydberg equation.

To satisfy the above conditions, we assumed that during excitation of a toryx its dimensions change as a function of the *toryx quantization parameter z* given by the equation:

$$z = 2(n\Lambda)^m \quad (m = 0, 1,..) , (n = 0, 1,..) \qquad (11.3\text{-}1)$$

where
m = toryx exponential excitation quantum state
n = toryx linear excitation quantum state
Λ = unified quantization constant.

Table 11.3.1. Exponential excitation quantum states m of toryces corresponding to different matter levels M.

Matter level	Exponential excitation quantum energy state m			
	Harmonic toryces	Excited a-toryces	Excited e-toryces	
$M = 1$	0	0	1	Light dark matter
$M = 2$	0	1	2	**Ordinary matter**
$M = 3$	0	2	3	Heavy dark matter
...	...	...	...	...

The toryx exponential excitation quantum state m depends on the matter level M as shown in Table 11.3.1. Notably, the matter level $M = 1$ corresponds to the light dark matter, the ordinary matter corresponds to the matter level $M = 2$, while the matter level M that is equal or greater than 3 corresponds to the heavy dark matter.

Table 11.3.2 shows quantization equations for relative radii of spherical membrane and leading string of basic excited toryces forming basic excited trons.

Table 11.3.2. Quantization equations for relative radii of spherical membranes and leading strings of basic toryces forming basic excited trons.

Tron names	Toryx	Relative radius of spherical membrane	Relative radius of toryx leading string
Ethertron $a^0_{m,n,q}$	$A^-_{m,n,q}$	$b^- = \dfrac{z+1}{z-1}$ (11.3-2a)	$b^-_1 = \dfrac{z}{z-1}$ (11.3-2b)
	$A^+_{m,n,q}$	$b^+ = \dfrac{z-1}{z+1}$ (11.3-3a)	$b^+_1 = \dfrac{z}{z+1}$ (11.3-3b)
Singulatron $\tilde{a}^0_{m,n,q}$	$\tilde{A}^-_{m,n,q}$	$\tilde{b}^- = \dfrac{1+z}{1-z}$ (11.3-4a)	$\tilde{b}^-_1 = \dfrac{1}{1-z}$ (11.3-4b)
	$\tilde{A}^+_{m,n,q}$	$\tilde{b}^+ = \dfrac{1-z}{1+z}$ (11.3-5a)	$\tilde{b}^+_1 = \dfrac{1}{1+z}$ (11.3-5b)
Electron $\overline{e}_{m,n,q}$	$E^-_{m,n,q}$	$b^- = 2z-1$ (11.3-6a)	$b^-_1 = z$ (11.3-6b)
	$\tilde{E}^-_{m,n,q}$	$\tilde{b}^- = 1-2z$ (11.3-7a)	$\tilde{b}^-_1 = 1-z$ (11.3-7b)
Positron $e^+_{m,n,q}$	$E^+_{m,n,q}$	$b^+ = \dfrac{1}{2z-1}$ (11.3-8a)	$b^+_1 = \dfrac{z}{2z-1}$ (11.3-8b)
	$\tilde{E}^+_{m,n,q}$	$\tilde{b}^+ = \dfrac{1}{1-2z}$ (11.3-9a)	$\tilde{b}^+_1 = \dfrac{1-z}{1-2z}$ (11.3-9b)

11.4 Harmonic Toryces

Harmonic toryces is a particular case of the toryces in which the toryx exponential excitation quantum state m approaches infinility (± 0).

Frequencies of trailing strings of these toryces relate to one another by simple harmonic ratios explaining their name. Since the exponential excitation quantum state m of harmonic toryces approaches infinility, their spacetime properties are the same for all matter levels. Table 11.4 shows relative radii of spherical membranes and leading strings of basic harmonic toryces forming basic harmonic trons.

Table 11.4. Relative radii of spherical membranes and leading strings of basic harmonic toryces.

Tron names	Toryx	Relative radius of spherical membrane	Relative radius of toryx leading string
Harmonic ethertron $a^0_{0,n,q}$	$A^{-\frac{1}{2}}_{0,n,q}$	$b^- = 3.0$	$b^-_1 = 2.0$
	$A^{+\frac{1}{2}}_{0,n,q}$	$b^+ = \frac{1}{3}$	$b^+_1 = \frac{2}{3}$
Harmonic singulatron $\breve{a}^0_{0,n,q}$	$\breve{A}^{-2}_{0,n,q}$	$\breve{b}^- = -3.0$	$\breve{b}^-_1 = -1.0$
	$\breve{A}^{+2}_{0,n,q}$	$\breve{b}^+ = -\frac{1}{3}$	$\breve{b}^+_1 = \frac{1}{3}$
Harmonic electron $e^-_{0,n,q}$	$E^{-\frac{1}{2}}_{0,n,q}$	$b^- = 3.0$	$b^-_1 = 2.0$
	$\breve{E}^{-2}_{0,n,q}$	$\breve{b}^- = -3.0$	$\breve{b}^-_1 = -1.0$
Harmonic positron $\breve{e}^+_{0,n,q}$	$E^{+\frac{1}{2}}_{0,n,q}$	$b^+ = \frac{1}{3}$	$b^+_1 = \frac{2}{3}$
	$\breve{E}^{+2}_{0,n,q}$	$\breve{b}^+ = -\frac{1}{3}$	$\breve{b}^+_1 = \frac{1}{3}$

Notably, for each basic a-harmonic toryx there is a basic e-harmonic toryx with the same spacetime properties, so:

$$A^{-\frac{1}{2}}_{0,n,q} = E^{-\frac{1}{2}}_{0,n,q} = AE^{-\frac{1}{2}}_{0,n,q}, \quad A^{+\frac{1}{2}}_{0,n,q} = E^{+\frac{1}{2}}_{0,n,q} = AE^{+\frac{1}{2}}_{0,n,q}$$
$$\breve{A}^{-2}_{0,n,q} = \breve{E}^{-2}_{0,n,q} = A\breve{E}^{-2}_{0,n,q}, \quad \breve{A}^{+2}_{0,n,q} = \breve{E}^{+2}_{0,n,q} = A\breve{E}^{+2}_{0,n,q} \qquad (11.4\text{-}1)$$

11.5 Self- and Mutually-Sustained Trons – Trons sustain their existence by absorption and release of energy. Depending on the way of performing this function they are divided into two groups called the *self-sustained trons* and the *mutually-sustained trons*.

Self-sustained trons - As we described in Chapter 8, some trons have an important feature when the relative radii of leading strings of their constituent toryces are within the ranges shown in Table 8.12.

$$0.544 < b_1 < 1.0$$
$$1.0 < b_1 < 6.222 \tag{11.5-1}$$

In Eq. (11.5-1), the top range corresponds to the real positive toryces while the bottom range corresponds to the real negative toryces. In both these toryces peripheral translational velocities of trailing strings periodically exceed speed of light, making rotational velocities imaginary. Consequently, these toryces periodically absorb and release energy, making their trons self-sustaining.

Mutually-sustained trons - In the mutually-sustained trons the relative radii of leading strings of their constituent toryces are outside the ranges specified by Eq. (11.5-1). These trons are made up out of real and imaginary toryces with the imaginary toryces responsible for absorption of energy and the real toryces responsible for releasing it. The amount of energy either released or absorbed by the mutually sustained toryces is equal to the exchange energy expressed by Eq. (10.4-5).

11.6 Oscillation of Toryces – During oscillation of a toryx, its radii, r_0 and r_1, change proportionally to the *toryx oscillation factor* Q_q as shown in Figure 11.6.1.

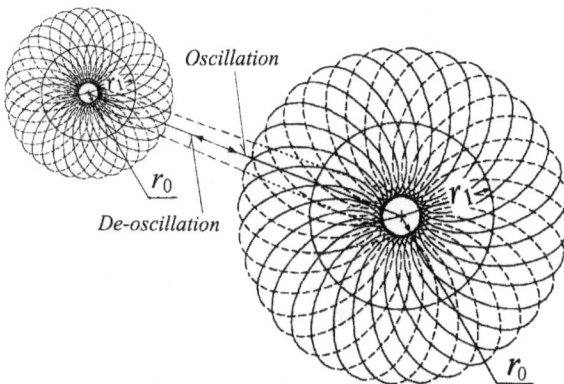

Figure 11.6.1. Oscillation and de-oscillation of a toryx.

To define the toryx oscillation factor Q_q, we assumed that the toryx oscillation factor Q_q is a function of the *toryx oscillation quantum state q* and the matter level constant Λ as given by the equation, allowing us to simulate quantum energy states of leptons:

$$Q_0 = 1$$

$$Q_q = 3\left(\frac{\Lambda}{2(q-1)}\right)^{q-1} \quad (q = 1, 2,..) \qquad (11.6\text{-}1)$$

The values of Q_q calculated from Eq. (11.6-1) are plotted in Figure 11.6.2 and some of them are shown Table 11.6.1. At $q = q_m$ and $\Lambda = 137$, the toryx oscillation factor Q_q reaches its maximum value Q_{qm}:

$$q_m = 1 + \frac{\Lambda}{2e} = 26.199742$$

$$Q_{qm} = 3e^{\Lambda/2e} = 2.637728 \times 10^{11} \qquad (11.6\text{-}2)$$

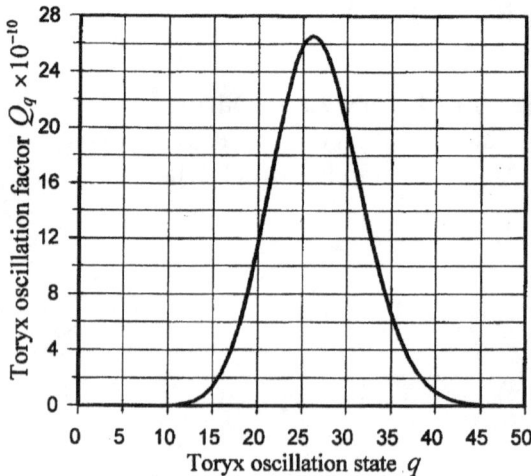

Figure 11.6.2. Toryx oscillation factor Q_q as a function of the toryx oscillation state q.

When $q > q_m$, the toryx oscillation factor Q_q sharply decreases and at $q = 70.589986$ its magnitude reduces to 1. After that as q continues to increase and approaches infinity, Q_q also decreases and approaches infinility.

Table 11.6.1. Toryx oscillation factor Q_q as a function of the toryx oscillation quantum state q.

q	0	1	2	3	4	5
Q_q	1.000	3.000	205.500	3519.1875	35713.236	258014.18

The relationship between the toryx parameters and the toryx oscillation factor Q_q is given by the equation:

$$Q_q = \frac{r_{i0}}{r_{iq}} = \frac{r_{10}}{r_{1q}} = \frac{f_{iq}}{f_{i0}} = \frac{m_{tgq}}{m_{tg0}} = \frac{m_{tiq}}{m_{ti0}} = \frac{\mu_{t0}}{\mu_{tq}} \qquad (11.6\text{-}3)$$

In Eq. (11.6-3), the parameters with the subscripts "0" and "q" correspond respectively to $q = 0$ and to $q > 0$. It follows from this equation that the toryx masses increase proportionally to Q_q, while its magnetic moment decreases inverse-proportionally to Q_q. Importantly, the toryx charge does not change during oscillation.

CHAPTER 12

BASIC MICRO-TRONS

We will describe in this Chapter structure and properties of three kinds of basic trons that form nucleons and atoms:

1. Quantum vacuum (QV) trons
2. Harmonic trons
3. Excited trons.

12.1 Quantum Vacuum – Quantum vacuum of this theory differs from the quantum vacuum of the quantum theory in several ways.

According to the quantum theory, quantum vacuum is a conglomerate of seething *virtual particles* produced by quantum fluctuations in according to the Heisenberg's Uncertainty Principle. In accordance with this principle, a virtual particle with the energy E may pop up into and out of existence spontaneously and will exist for the period of time ΔT providing that the product of the particle energy E by the period of time ΔT does not exceed the ratio of the Planck constant h over 2π, as given by the equation:

$$E \cdot \Delta T \leq \frac{h}{2\pi} \qquad (12.1\text{-}1)$$

In quantum theory, virtual particles are not real and can never be detected directly. Their spacetime properties are not identified. The key role of virtual particles is to determine the way how "real" particle interact. Since the virtual particles are not real, they do not have to comply with the conservation laws, which are usually obeyed during the interactions of stable particles.

According to the UST, quantum vacuum (Fig. 12.1) is made up of ethertrons, electrons, positrons a singulatrons at the lowest linear excitation quantum state $n = 0$. Based on Eqs. (11.3-1), (11.3-2b) – (11.3-9b), in the real toryces of the quantum vacuum the relative radii of leading and trailing strings are: $b_1 = 0$ and $b_2 = 1$, while in the imaginary toryces they are: $b_1 = 1$ and $b_2 = 0$. To obey the Unified Law of Stable Polarization, the relationship between the number of real and imaginary toryces of the quantum is defined by Eq. (11.2-7).

Table 12.1. Components and calculated physical properties of QV-trons and their constituent toryces forming the quantum vacuum.

QV-trons	Constituent toryces					
	Symbol	b_1	b_2	e/e_0	m_g/m_e	B_t
Ethertron $a^0_{m,0,q}$	$A^-_{m,0,q}$	- 0	- 1	$-\infty$	∞	$\pm\infty$
	$A^+_{m,0,q}$	+ 0	- 1	$+\infty$	∞	$\pm\infty$
Singulatron $\breve{a}^0_{m,0,q}$	$\breve{A}^-_{m,0,q}$	1	+ 0	- 0	0	+ 0
	$\breve{A}^+_{m,0,q}$	1	- 0	+ 0	0	- 0
Electron $e^{-1}_{m,0,q}$	$E^-_{m,0,q}$	+ 0	- 1	$+\infty$	∞	$-\infty$
	$\breve{E}^-_{m,0,q}$	1	+ 0	+ 0	0	- 0
Positron $e^{+1}_{m,0,q}$	$E^+_{m,0,q}$	- 0	- 1	$-\infty$	∞	$+\infty$
	$\breve{E}^+_{m,0,q}$	1	+ 0	- 0	0	+ 0

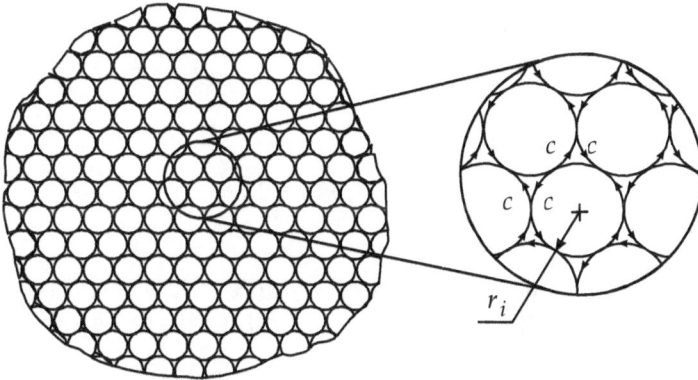

Figure 12.1. Quantum vacuum of the UST.

Notably, at the lowest linear excitation quantum state $n = 0$, the real toryces become imaginary and the imaginary toryces become real. Consequently, the quantum vacuum consists of two parts, a central part and a peripheral part. The central part is made up of the following real toryces with the relative radii of leading and trailing strings $b_1 = 0$ and $b_2 = 1$:

$$A^{-}_{m,0,q}, A^{+}_{m,0,q}, E^{-}_{m,0,q}, E^{-}_{m,0,q}, E^{+}_{m,0,q},$$

The peripheral part is made up of infinite numbers of real toryces with the relative radii of leading and trailing strings $b_1 = 1$ and $b_2 = 0$:

$$\breve{A}^{-}_{m,0,q}, \breve{A}^{+}_{m,0,q}, \breve{E}^{-}_{m,0,q}, \breve{E}^{-}_{m,0,q}, \breve{E}^{+}_{m,0,q}$$

12.2 Creation of Harmonic Trons – According to our theory, the process of creation of harmonic toryces from quantum vacuum is governed by the *Unified Uncertainty Principle* that is expressed by the equation:

$$abs(E_m T_1) \le \frac{h}{4\pi} \tag{12.2-1}$$

In the above equation E_m is the toryx matter energy and T_1 is the period of toryx leading string. This equation yields two ranges of relative radii of leading strings b_{1h} of the harmonic toryces that may pop into and out of existence spontaneously.

$$abs\left(\frac{(b_{1h}-1)b_{1h}}{\sqrt{2b_{1h}-1}}\right) \le \frac{1}{\pi\alpha} \quad (0.500016 < b_{1h} < 16.12296)$$

$$abs\left(\frac{(b_{1h}-1)b_{1h}}{i\sqrt{2b_{1h}-1}}\right) \le \frac{1}{\pi\alpha} \quad (0.499984 > b_{1h} > -15.12296) \tag{12.2-2}$$

Described below are the structures and properties of basic harmonic trons with the relative radii of leading strings b_{1h} within the ranges described by Eq. (12.2-2).

12.3 Basic Harmonic Ethertrons - Each basic harmonic ethertron $a^0_{0,n,0}$ is made up of the basic real negative harmonic toryx $A^{-\frac{1}{2}}_{0,n,q}$ and the basic real positive harmonic toryx $A^{+\frac{1}{2}}_{0,n,q}$ that coexist intermittently according to the equation:

$$a^0_{0,n,0} = A^{-\frac{1}{2}}_{0,n,q} + A^{+\frac{1}{2}}_{0,n,q} \tag{12.3-1}$$

Figure 12.3 shows cross-section of the basic harmonic ethertron $a_{0,n,0}^{0}$ and spacetime parameters of its constituent basic real negative harmonic toryx $A_{0,n,q}^{-\frac{1}{2}}$ and the basic real positive harmonic toryx $A_{0,n,q}^{+\frac{1}{2}}$. Quantum states of the constituent toryces are defined by Eqs. (11.3-2) and (11.3-3). Table 12.3 shows components and physical properties of the basic harmonic ethertron.

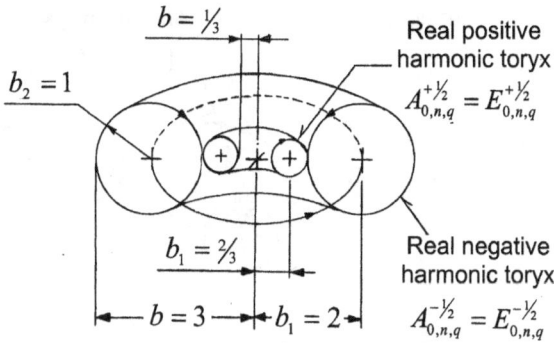

Figure 12.3. Cross-section and spacetime parameters of basic real harmonic toryces making up basic harmonic ethertron.

Table 12.3. Components and physical properties of basic harmonic ethertron ($m = 0$, $n \geq 1$, $q = 0$).

Toryx	b_1	e_t/e	μ_t/μ_N	m_{tg}/m_e	P_R	N
$A_{0,n,0}^{-\frac{1}{2}}$	2.0	- 0.50	∓ 5.80115555	0.500000	$+\frac{3}{6}$	1
$A_{0,n,0}^{+\frac{1}{2}}$	$\frac{2}{3}$	+ 0.50	± 1.93371852	0.500000	$+\frac{1}{6}$	1
Harmonic ethertron $a_{0,n,0}^{0}$	± 0	∓ 1.93371852	**0.500000**	$+\frac{5}{6}$	**1**

12.4 Basic Harmonic Singulatrons – Each basic harmonic singulatron $\breve{a}_{0,n,0}^{0}$ is made up of the basic imaginary negative harmonic toryx $\breve{A}_{0,n,q}^{-2}$ and the basic imaginary positive harmonic toryx $\breve{A}_{0,n,q}^{+2}$ that coexist intermittently according to the equation:

$$\breve{a}_{0,n,0}^{0} = \breve{A}_{0,n,q}^{-2} + \breve{A}_{0,n,q}^{+2} \qquad (12.4\text{-}1)$$

Figure 12.4 and Table 12.4 show a cross-section of the basic harmonic singulatron $\breve{a}_{0,n,0}^{0}$ and spacetime parameters of its constituent basic imaginary negative harmonic toryx $\breve{A}_{0,n,q}^{-2}$ and basic imaginary positive harmonic toryx $\breve{A}_{0,n,q}^{+2}$. Quantum states of the constituent basic toryces are defined by Eqs. (11.3-4) and (11.3-5).

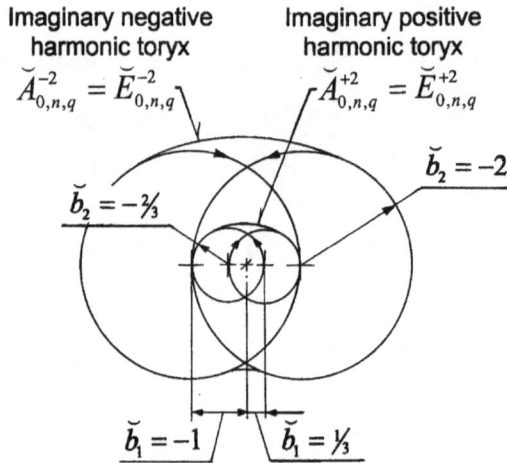

Imaginary negative harmonic toryx
$$\breve{A}_{0,n,q}^{-2} = \breve{E}_{0,n,q}^{-2}$$

Imaginary positive harmonic toryx
$$\breve{A}_{0,n,q}^{+2} = \breve{E}_{0,n,q}^{+2}$$

$$\breve{b}_2 = -2$$
$$\breve{b}_2 = -\tfrac{2}{3}$$
$$\breve{b}_1 = -1$$
$$\breve{b}_1 = \tfrac{1}{3}$$

Figure 12.4. Cross-section and spacetime parameters of basic imaginary harmonic toryces making up basic harmonic singulatron.

Table 12.4. Components and physical properties of basic harmonic singulatron ($m = 0, n \geq 1, q = 0$).

Toryx	b_1	e_t/e	μ_t/μ_N	m_{tg}/m_e	P_R	N
$\breve{A}_{0,n,0}^{-2}$	-1	- 2.0	$\mp$ 23.20462221	2.000000	-6.0	1
$\breve{A}_{0,n,0}^{+2}$	$\tfrac{1}{3}$	+ 2.0	$\pm$ 7.73487407	2.000000	$-\tfrac{2}{3}$	1
Harmonic singulatron $\breve{a}_{0,n,0}^{0}$	± 0		$\mp$ 7.73487407	**2.000000**	$-\tfrac{10}{3}$	**1**

12.5 Basic Harmonic Electrons

– Each basic harmonic electron $e_{0,n,0}^{-5/4}$ is made up of one basic imaginary negative harmonic toryx $\breve{E}_{0,n,q}^{-2}$ and the basic real negative harmonic toryces $E_{0,n,q}^{-1/2}$ that coexist intermittently according to the equation:

$$e_{0,n,q}^{-5/4} = \breve{E}_{0,n,q}^{-2} + E_{0,n,q}^{-1/2} \qquad (12.5\text{-}1)$$

Table 12.5 shows components and physical properties of the basic harmonic electron $e_{0,n,q}^{-2}$ and spacetime parameters of its constituent basic imaginary negative harmonic toryx $\breve{E}_{0n,q}^{-2}$ and basic real negative harmonic toryx $E_{0,n,q}^{-1/2}$. Quantum states of this electron are defined by Eqs. (11.3-6) and (11.3-7).

Table 12.5. Components and physical properties of basic harmonic electron ($m = 0$, $q = 0$).

Toryx	b_1	e_t/e	μ_t/μ_N	m_{tg}/m_e	P_R	N
$\breve{E}_{0,n,0}^{-2}$	- 1.0	- 2.00	∓ 23.20462221	2.000000	- 6.00	1
$E_{0,n,0}^{-1/2}$	2.0	- 0.50	∓ 5.80115555	0.500000	+ 1.50	1
Harmonic electron $e_{0,n,0}^{-5/4}$		- 1.25	∓ 14.50288888	1.250000	- 2.25	**1**

12.6 Basic Harmonic Positron

12.6 Basic Harmonic Positron – Each basic harmonic positron $\breve{e}_{0,n,0}^{+5/4}$ is made up of one basic imaginary positive harmonic toryx $\breve{E}_{0,n,q}^{+2}$ and one basic real positive harmonic toryx $E_{0,n,q}^{+1/2}$ that coexist intermittently according to the equation:

$$e_{0,n,q}^{+5/4} = \breve{E}_{0,n,q}^{+2} + E_{0,n,q}^{+1/2} \qquad (12.6\text{-}1)$$

Table 12.6. Components and physical properties of basic harmonic positron ($m = 0$, $q = 0$).

Toryx	b_1	e_t/e	μ_t/μ_N	m_{tg}/m_e	P_R	N
$\breve{E}_{0,n,q}^{+2}$	$1/3$	+ 2.00	± 7.73487407	2.000000	- $2/3$	1
$E_{0,n,q}^{+1/2}$	$2/3$	+ 0.50	± 1.93371852	0.500000	+ $1/6$	1
Harmonic positron $e_{0,n,0}^{+5/4}$		+ 1.25	± 4.83429629	1.250000	- 0.25	**1**

Table 12.6 shows components and physical properties of the basic imaginary harmonic positron $e_{0n,0}^{+5/4}$ and spacetime parameters of its constituent basic imaginary positive harmonic toryx $\breve{E}_{0,nq}^{+2}$ and basic real harmonic toryx $E_{0nq}^{+1/2}$. Quantum states of this positron are defined by Eqs. (11.3-8) and (11.3-9).

12.7 Formation of Excited Trons – An excited tron is formed after its constituent harmonic toryces becomes excited. In the excited trons and their constituent toryces, the exponential quantum energy state m is equal to or greater than 1.

According to our theory, the process of formation of excited toryces from harmonic toryces is possible when the following condition is satisfied:

$$abs(E_{xe}\,T_{1e} - E_{xh}\,T_{1h}) \leq \frac{h}{2} \qquad (12.7\text{-}1)$$

where
E_{xe} = exchange energy of excited toryx
E_{xh} = exchange energy of harmonic toryx
T_{1e} = period of leading string of excited toryx
T_{1h} = period of leading string of harmonic toryx.

Let b_{1h} to be the relative radius of leading string of a harmonic toryx and b_{1e} the relative radius of leading string of an excited toryx. Then, based on Eq. (12.7-1), the excited toryces will be formed from harmonic toryces when the relationships between relative radii of leading strings b_{1h} and b_{1e} satisfy the following condition:

$$abs\left(\frac{(b_{1e}-1)\sqrt{2b_{1e}-1}}{b_{1e}} - \frac{(b_{1h}-1)\sqrt{2b_{1h}-1}}{b_{1h}}\right) \leq \frac{4}{\alpha}$$

$$abs\left(\frac{i(b_{1e}-1)\sqrt{2b_{1e}-1}}{b_{1e}} - \frac{i(b_{1h}-1)\sqrt{2b_{1h}-1}}{b_{1h}}\right) \leq \frac{4}{\alpha} \qquad (12.7\text{-}2)$$

12.8 Basic Excited Ethertrons –

Each basic excited ethertron $a^0_{m,n,q}$ is made up of the basic real negative excited a-toryx $A^-_{m,n,q}$ and the basic real positive excited a-toryx $A^+_{m,n,q}$ that coexist intermittently according to the equation:

$$a^0_{m,nq} = A^-_{m,n,q} + A^+_{m,n,q} \qquad (12.8\text{-}1)$$

Figure 12.8 shows a cross-section and spacetime parameters of the basic excited ethertron $a^0_{1,1,q}$ of the ordinary matter $M = 2$.

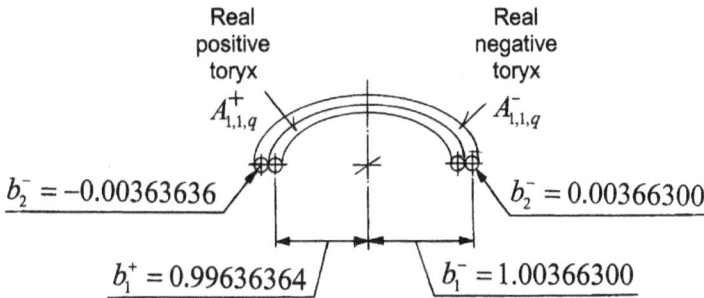

Figure 12.8. Cross-section and spacetime parameters of basic excited ethertron $a^0_{1,1,q}$ of the ordinary matter $M = 2$. ($m = 1$, $n = 1$).

Table 12.8 shows components and physical properties of the basic excited ethertrons $a^0_{1,1,0}$ of ordinary matter $M = 2$. Quantum states of its constituent toryces are defined by Eqs. (11.3-2) and (11.3-3).

Table 12.8. Components and physical properties of the basic excited ethertron of the ordinary matter $M = 2$. ($m = 1$, $n = 1$, $q = 0$).

Toryx	b_1	e_t / e	μ_t / μ_N	P_R	m_{tg} / m_e
$A^-_{1,1,0}$	1.0036630	- 0.0036496	∓ 0.0245368	+ 0.0036764	0.00364964
$A^+_{1,1,0}$	0.9963636	+ 0.0036496	± 0.0243584	+ 0.0036231	0.00364964
Excited ethertron $a^0_{1,1,0}$	0.0000000		∓ 0.0000892	+0.0036497	0.00364964

12.9 Basic Excited Singulatrons – Each basic excited singulartron $\breve{a}^0_{m,n,q}$ is made up of the basic imaginary negative excited a-toryx $\breve{A}^-_{m,n,q}$ and the basic imaginary positive excited a-toryx $\breve{A}^+_{m,n,q}$ that coexist intermittently according to the equation:

$$\breve{a}^0_{m,n,q} = \breve{A}^-_{m,n,q} + \breve{A}^+_{m,n,q} \qquad (12.9\text{-}1)$$

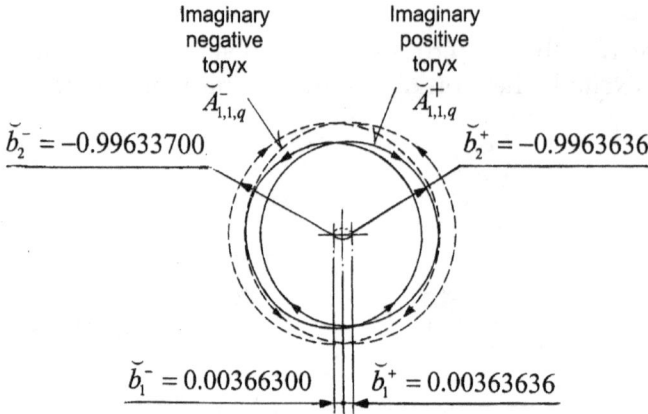

Figure 12.9. Cross-section and spacetime parameters of basic excited singulatron $\breve{a}^0_{1,1,q}$ of the ordinary matter $M = 2$. ($m = 1$, $n = 1$).

Figure 12.9 shows a cross-section and spacetime parameters of the basic excited singulatron $\breve{a}^0_{1,1,q}$ of ordinary matter $M = 2$. Table 12.9 shows components and physical properties of this singulatron. Quantum states of its constituent toryces are defined by Eqs. (11.3-4) and (11.3-5).

Table 12.9. Components and physical properties of basic excited singulatron of the ordinary matter $M = 2$. ($m = 1$, $n = 1$, $q = 0$).

Toryx	b_1	e_t / e	μ_t / μ_N	P_R	m_{tg} / m_e
$\breve{A}^-_{1,1,0}$	- 0.0036630	-274.0	∓ 1842.126564	- 276.007326	274.0000
$\breve{A}^+_{1,1,0}$	0.0036364	+274.0	± 1828.729280	- 272.007273	274.0000
Excited singulatron $\breve{a}^0_{2,1,0}$	**0.0**		$\mp\, 6.69864205$	- 274.007299	**274.0000**

12.10 Basic Excited Electrons – Each basic excited electron $e^{-1}_{m,n,q}$ is made up of one real basic negative excited toryx $E^{-}_{m n q}$ and one imaginary negative excited toryx $\breve{E}^{-}_{m n q}$ that coexist intermittently according to the equation:

$$e^{-1}_{m,n,q} = E^{-}_{m,n,q} + \breve{E}^{-}_{m,n,q} \qquad (12.10\text{-}1)$$

Table 12.10 shows components and physical properties of this electron. Quantum states of the relative radii of its constituent toryces are defined by Eqs. (11.3-6) and (11.3-7).

Table 12.10. Components and physical properties of excited electron of the ordinary matter $M = 2$. ($m = 2$, $n = 1$, $q = 0$).

Toryx	b_1	e_t / e	μ_t / μ_N	m_{tg} / m_e	P_R	N
$E^{-}_{2,1,0}$	37538	- 0.9999734	- 1835.35458	0.9999734	+ 75073	1
$\breve{E}^{-}_{2,1,0}$	- 37537	- 1.0000266	- 1835.45237	1.0000266	- 75077	1
Excited electron $e^{-1}_{2,1,0}$		- 1.0000000	- 1835.403475	1.0000000	-2.0	1

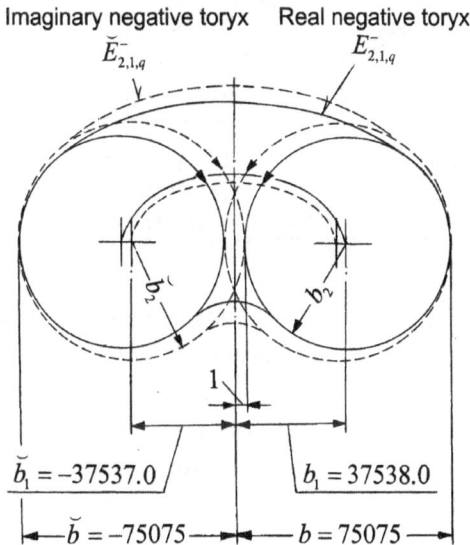

Figure 12.10. Cross-section and spacetime parameters of basic excited electron $e^{-1}_{2,1,q}$ of the ordinary matter $M = 2$. ($m = 2$, $n = 1$).

Figure 12.10 shows a cross-section and spacetime parameters of the basic excited electron $e_{2,1,q}^{-1}$ of the ordinary matter $M = 2$.

12.11 Basic Excited Positrons - Each basic excited positron $e_{m,n,q}^{+1}$ is made up of one basic real positive excited toryx $E_{m,n,q}^{+}$ and one basic imaginary positive excited toryx $\breve{E}_{m,n,q}^{+}$ that coexist intermittently according to the equation:

$$e_{m,n,q}^{+1} = E_{m,n,q}^{+} + \breve{E}_{m,n,q}^{+} \qquad (12.11\text{-}1)$$

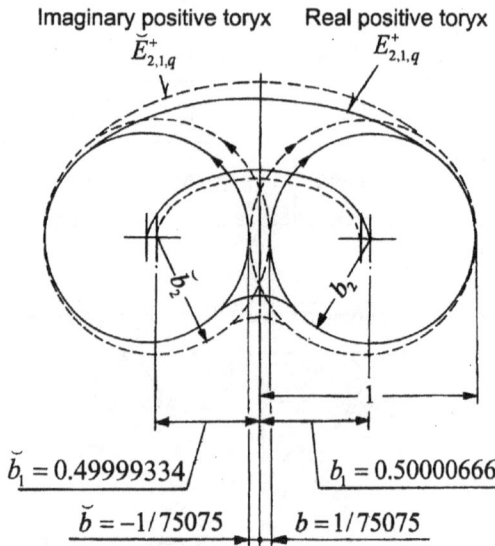

Imaginary positive toryx $\breve{E}_{2,1,q}^{+}$ Real positive toryx $E_{2,1,q}^{+}$

$\breve{b}_1 = 0.49999334$ $b_1 = 0.50000666$

$\breve{b} = -1/75075$ $b = 1/75075$

Figure 12.11. Cross-section and spacetime parameters of the excited positron $e_{2,1,q}^{+1}$ of the ordinary matter $M = 2$. ($m = 2$, $n = 1$).

Table 12.11 shows components and physical properties of this positron. Quantum states of its constituent toryces are defined by Eqs. (11.3-8) and (11.3-9).

Figure 12.11 shows a cross-section and spacetime parameters the basic excited positron $e_{2,1,q}^{+1}$ of the ordinary matter $M = 2$.

Table 12.11. Components and physical properties of excited positron of the ordinary matter $M = 2$. ($m = 2, n = 1, q = 0$).

Toryx	b_1	e_t / e	μ_t / μ_N	m_{tg} / m_e	P_R	N
$E_{2,1,0}^+$	0.50000666	+ 0.9999734	+ 0.02444695	0.9999734	$+10^{-5}$	1
$\breve{E}_{2,1,0}^+$	0.49999334	+ 1.0000266	+ 0.02444825	1.0000266	-10^{-5}	1
Excited positron $e_{2,1,0}^{+1}$		+ **1.0000000**	+ **0.02444760**	**1.0000000**	**0.00**	**1**

CHAPTER 13

BASIC HYDROGEN ATOM

In the previous two chapters we described the formation of basic trons. In this chapter we describe the formation of a basic hydrogen atom. The basic hydrogen atom is made up of two main parts, *nucleon core* and *atomic shell*.

The main components of the nucleon core are:

- Nucleon octahedron
- Harmonic a-bond
- Excited a-bond.

The main components of the atomic shell are:

- Excited electron
- Excited positron.

13.1 Nucleon Octahedron – According to our theory, for the trons to form hydrogen atoms and nucleons their degrees of freedom must be substantially reduced. This is accomplished by locating the trons at the center and vertices of an octahedral crystal structure called the *nucleon octahedron.*

The nucleon octahedron is formed by six self-sustained basic real harmonic ethertrons $a_{0,n,q}^0$ with the structure of each ethertron given by the equation:

$$a_{0,n,0}^0 = A_{0,n,0}^{-\frac{1}{2}} + A_{0,n,0}^{+\frac{1}{2}} \qquad (13.1\text{-}1)$$

Figure 12.3 and Table 12.3 show structures and properties of these toryces. As shown in Figure 13.1.1, the nucleon octahedron $0_{0,n,q}^0$ has a dual octahedral structure containing outer and inner octahedrons. The outer octahedron is formed by the negative harmonic toryces $A_{0,n,q}^{-\frac{1}{2}}$ and the inner octahedron is formed by the positive harmonic toryces $A_{0,n,q}^{+\frac{1}{2}}$. These toryces coexist intermittently.

In both outer and inner octahedron, the nucleon center 1 is located at the center 1 of six vertices 2 through 7. In the outer octahedron, horizontal and vertical distances between adjacent vertices are equal to the diameter of the real inversion toryx $2r_i$. In the inner octahedron, these distances are three times shorter.

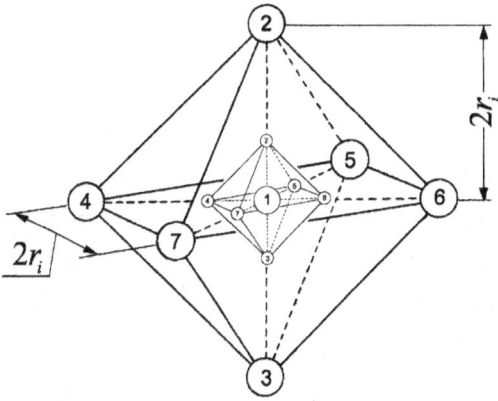

Fig. 13.1.1. Nucleon octahedron $O_{0,n,q}^0$.

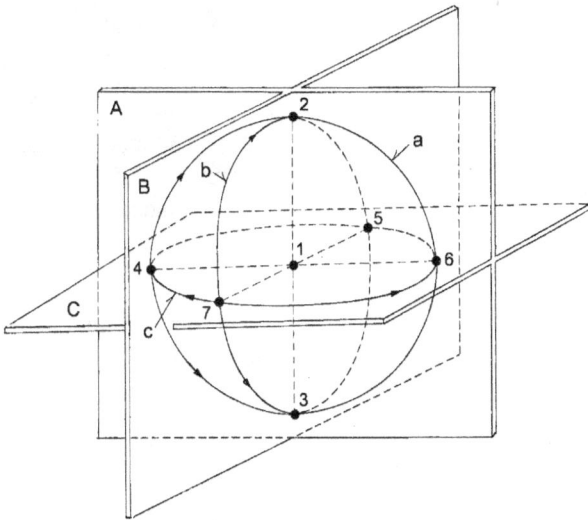

Fig. 13.1.2. Formation of vertices and center of nucleon octahedron.

Figure 13.1.2 shows the space orientations of the self-sustained negative harmonic toryces $A_{0,n,q}^{-1/2}$. The leading strings a, b and c of these toryces reside in three mutually-perpendicular planes A, B and C. The centers of the harmonic toryces are located at the intersection of these planes that also coincides with location of the nucleon center 1. Consequently, the leading strings a, b and c of the negative harmonic toryces $A_{0,n,q}^{-1/2}$ appear as three *outer great circles*. The vertices 2 through 7 of the outer octahedron are formed at the intersections of leading strings a, b and c of the negative harmonic toryces.

In the inner octahedron the vertices are formed at the intersections of leading strings of self-sustained positive harmonic toryces $A_{0,n,q}^{+1/2}$ residing in three mutually-perpendicular planes A, B and C. Consequently, the leading strings a, b and c of the positive harmonic toryces $A_{0,n,q}^{+1/2}$ appear as three *inner great circles*. The vertices 2 through 7 of the inner octahedron are formed at the intersections of leading strings a, b and c of the positive harmonic toryces.

Table 13.1. Locations and properties of components of nucleon octahedron.

Tron – Location (Fig. 13.1.2)	μ / μ_N	m_g / m_e	P_R	N
$a_{0,n,0}^{0}$ - Plane A	± 0.00000000	0.500000	$+ \frac{5}{6}$	2
$a_{0,n,0}^{0}$ - Plane B	± 0.00000000	0.500000	$+ \frac{5}{6}$	2
$a_{0,n,0}^{0}$ - Plane C	± 0.00000000	0.500000	$+ \frac{5}{6}$	2
Nucleon octahedron $O_{0,n,0}^{0}$	± 0.00000000	1.500000	$+ 2.50$	-

Table 13.1 shows locations and properties of components of the nucleon octahedron $O_{0,n,q}^{0}$. A pair of harmonic trons is located in each of three planes A, B and C. In each pair the magnetic moments of its harmonic trons have opposite polarities, reducing the magnetic moment of the pair to zero.

13.2 Harmonic A-Bond – The harmonic a-bond $ab_{0,n,q}^{0}$ is assembled from one basic imaginary harmonic singulatrons $\breve{a}_{0,n,q}^{0}$ and two pairs of basic real harmonic ethertrons $a_{0,n,q}^{0}$ with opposite polarities of their magnetic moments according to the equation:

$$ab_{0,n,q}^{0} = \downarrow \breve{a}_{0,n,q}^{0} + 2(\uparrow a_{0,n,q}^{0} + \downarrow a_{0,n,q}^{0}) \qquad (13.2\text{-}1)$$

Cross-sections and spacetime parameters of these trons are shown in Figures 12.3 and 12.4; their components and physical properties are shown in Tables 12.3 and 12.4. Harmonic singulatrons and ethertrons are mutually-sustained. Therefore, according to Eq. (11.2-7), each harmonic singulatron must be matched with four harmonic ethertrons to obey the Unified Law of Stable Polarization.

Figure 13.2 shows locations of components of the harmonic a-bond $ab_{0,n,q}^0$ inside the nucleon octahedron. The singulatron $\downarrow \tilde{a}_{0,n,q}^0$ resides at the nucleon center 1, while two pairs of ethertrons $\downarrow a_{0,n,0}^0$ and $\uparrow a_{0,n,0}^0$ reside intermittently at the vertices 4 – 7 of the nucleon octahedron.

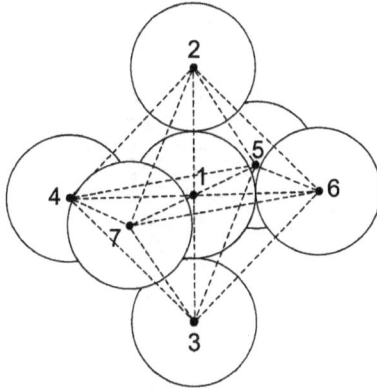

Fig. 13.2. Locations of components of harmonic a-bond.

Table 13.2. Components and physical properties of harmonic a-bond ($T = 4$).

Trons	b_1^-	μ / μ_N	m_g / m_e	P_R	N	Fig. 13.2
$\downarrow \tilde{a}_{0,n,0}^0$	- 1.00	- 7.73487407	2.000000	- $^{10}\!/_3$	1	1
$\uparrow a_{0,n,0}^0$	$^1\!/_3$	+ 3.86743704	0.500000	+ $^5\!/_6$	2	4, 6
$\downarrow a_{0,n,0}^0$	$^1\!/_3$	- 3.86743704	0.500000	+ $^5\!/_6$	2	5, 7
Harmonic a-bond $ab_{0,n,0}^0$		- 3.86743704	2.000000	± 0	-	-

Table 13.2 shows components and physical properties of the harmonic a-bond $ab_{0,n,0}^0$. Notably, the total reality polarization of the harmonic a-bond $ab_{0,n,0}^0$ is reduced to zero. In the a-bond, the imaginary singulatron coexists intermittently with its four real ethertrons.

13.3 Excited A-Bond – The excited a-bond $ab_{m,n,q}^0$ is assembled from the basic imaginary excited singulatrons $\breve{a}_{m,n,q}^0$ and the basic real excited ethertrons $a_{m,n,q}^0$ according to the equation:

$$ab_{m,n,q}^0 = 3(\downarrow \breve{a}_{m,n,q}^0 + T \downarrow a_{m,n,q}^0) + 4(\uparrow \breve{a}_{m,n,q}^0 + T \uparrow a_{m,n,q}^0) \quad (13.3\text{-}1)$$

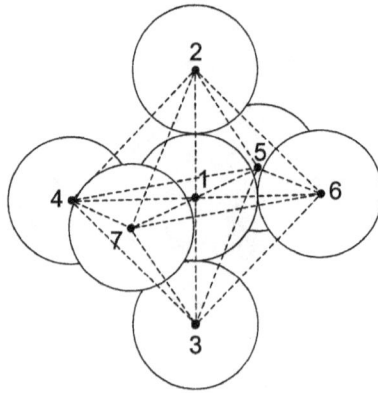

Fig. 13.3. Locations of components of excited a-bond.
1 - 3: singulatrons and ethertrons with negative magnetic moments
4 – 7: singulatrons and ethertrons with positive magnetic moments

Cross-sections and spacetime parameters of these trons are shown in Figures 12.8 and 12.9; their components and physical properties are shown in Tables 12.8 and 12.9. Excited singulatrons and ethertrons are mutually-sustained. Therefore, according to Eq. (11.2-7), each excited singulatron must be matched with T ethertrons to obey the Unified Law of Stable Polarization.

Figure 13.3 shows locations of components of the excited a-bond $ab_{m,n,q}^0$ inside the nucleon octahedron. Three singulatrons $\downarrow \breve{a}_{m,n,q}^0$ and 3T ethertrons $\downarrow a_{m,n,q}^0$ reside at the vertices 1 – 3 while four singulatrons $\uparrow \breve{a}_{2,1,q}^0$ and 4T ethertrons $\uparrow a_{m,n,q}^0$ reside at the vertices 4 – 7 of the nucleon octahedron. The imaginary singulatrons coexist intermittently with the real ethertrons.

Table 13.3. Components and physical properties of excited a-bond of the ordinary matter level $M = 2$ ($m = 1$).

Trons	b_1^-	μ / μ_N	m_g / m_e	P_R	N	Fig. 13.4
$\downarrow \tilde{a}_{1,1,0}^0$	- 0.0036630	- 6.69864205	274.000	-274.007	3	1–3
$\uparrow \tilde{a}_{1,1,0}^0$	0.0036364	+ 6.69864205	274.000	-274.007	4	4–7
$\downarrow a_{1,1,0}^0$	- 0.0036630	- 6.69864205	274.000	-274.007	3T	1–3
$\uparrow a_{1,1,0}^0$	0.0036364	+ 6.69864205	274.000	-274.007	4T	4–7
Excited a-bond $ab_{1,1,0}^0$		**+ 6.69864205**	**1918.000**	**± 0.00**	-	-

Table 13.3 shows components and physical properties of the basic nucleon core of the ordinary matter level $M = 2$. Notably, the total magnitude of absolute values of the reality polarizations P_R of all singulatrons is the same as the total magnitude of absolute values of the reality polarizations P_R of all ethertrons, but their signs are opposite, reducing the total reality polarization of the excited a-bond $ab_{1,1,0}^0$ to zero.

13.4 Nucleon Core – The nucleon core $nc_{m,n,q}^0$ is made up of two components: the harmonic bond $ab_{0,n,q}^0$ and the excited a-bond $ab_{m,n,q}^0$

Table 13.4 shows components and physical properties of the nucleon core $nc_{1,1,0}^0$ of the ordinary matter level $M = 2$.

Table 13.4. Components and physical properties of nucleon core of the ordinary matter level $M = 2$.

Nucleon core and its components	Symbols	μ / μ_N	m_g / m_e	P_R
Harmonic a-bond	$ab_{0,n,q}^0$	- 3.86743704	3.50	+ 2.50
Excited a-bond	$ab_{1,1,0}^0$	+ 6.69864205	1918.00	0.00
Nucleon core	$nc_{1,1,0}^0$	**+ 2.83120502**	**1921.50**	**+ 2.50**

13.5 Basic Hydrogen Atom – The basic hydrogen atom $^1H_{m,n,q}^0$ is made up of two components: the nucleon core $nc_{m,n,q}^0$, the excited electron $e_{m,n,q}^{-1}$ and the excited positron $e_{m,n,q}^{+1}$.

The structures of the excited electron $e_{2,1,0}^{-1}$ and the excited positron $e_{2,1,0}^{+1}$ of ordinary matter are respectively shown in Figures 12.10 and 12.11; their components and properties are shown in Tables 12.10 and 12.11.

Table 13.5 shows components and physical properties of the basic hydrogen atom $^1H_{m,n,q}^0$ of the ordinary matter level $M = 2$.

Table 13.5. Components and physical properties of basic hydrogen atom of the ordinary matter level $M = 2$.

Basic hydrogen atom and its components	Symbols	μ / μ_N	m_g / m_e	P_R
Nucleon core	$nc_{1,1,0}^0$	+ 2.83120502	1921.50	+ 2.50
Excited electron	$e_{2,1,0}^{-1}$	- 1835.403475	1.00	-2.00
Excited positron	$e_{2,1,0}^{+1}$	+ 0.02444760	1.00	0.00
Basic hydrogen atom	$^1H_{2,1,0}^0$	- 1832.547823	1923.50	+ 0.50

13.6 Hydrogen Atoms of Various Matter Levels – Both structure and properties of the basic hydrogen atom are greatly dependent on the matter level M.

Table 13.6. Components and properties of basic hydrogen atoms of the matter levels $M = 1$, 2 and 3.

Matter level	Basic hydrogen atoms	$\breve{b}_1^-$	$\breve{b}_1^+$	μ / μ_N	m_g / m_e
$M = 1$	$^1H_{1,1,0}^0$	274.00	0.50091410	- 152.514202	19.500
$M = 2$	$^1H_{2,1,0}^0$	37538.00	0.50000667	- 1832.54782	1923.50
$M = 3$	$^1H_{3,1,0}^0$	5142706.0	0.50000000	- 21480.1557	262771.5

Table 13.6 shows components and physical properties of the basic hydrogen atoms $^1H^0_{m,n,q}$ of the light dark matter level $M = 1$, the ordinary matter level $M = 2$ and the heavy dark matter level $M = 3$.

Since the toryx exponential quantum state m increases with increase of the matter level M, the toryx excitation quantization parameter z increases according to Eq. (11.3-1). Consequently, this produces substantial change in both structure and the gravitational mass of the basic hydrogen atom $^1H^0_{m,n,q}$.

- Increase of size of hydrogen atom - As the toryx quantization parameter z increases, the relative radii b_1^- and b_1^- of leading strings of real and imaginary negative toryces of the excited electron $e^{-1}_{m,n,q}$ increase according to Eqs. (11.3-6b) and (11.3-6b). Consequently, the size of the hydrogen atom increases.

- Increase of gravitational mass of hydrogen atom - As the toryx quantization parameter z increases, the relative radii b_1^- and b_1^+ of leading strings of imaginary negative and positive toryces of the excited singulatron $\tilde{a}^0_{m,n,q}$ decrease according to Eqs. (11.3-4b) and (11.3-5b). Consequently, the gravitational masses of the excited singulatron $\tilde{a}^0_{m,n,q}$ increase according to Eq. (10.2-2), leading to the increase of the gravitational mass of the basic nucleon core.

It follows from Table 13.6:

- Gravitational mass of the basic hydrogen atom of ordinary matter level $M = 2$ is approximately 99 times greater than the gravitational mass of the basic nucleon core of the light dark matter level $M = 1$.
- Gravitational mass of the basic hydrogen atom of the heavy dark matter $M = 3$ is approximately 136.6 times greater than the gravitational mass of the basic nucleon core of the ordinary matter $M = 2$.

CHAPTER 14

DERIVATIVE ATOMS, ISOTOPES & MOLECULES

In the previous chapter we described a way of assembling a basic hydrogen atom and nucleons from basic trons. In this chapter we describe structure and properties of several derivatives of the basic hydrogen atom. We show that all nucleons have the same nucleon core as the nucleon core of the basic hydrogen atom. The differences between the nucleons are in the structures of their shells. The basic hydrogen atoms and its derivatives form jointly many known atomic structures.

14.1 Quantum States of Derivative Hydrogen Atoms –
According to the UST, to form derivatives of the basic hydrogen atom it is necessary to modify quantum states of its electrons and positrons.

Consequently, the imaginary e-toryces have one additional quantum state called the toryx *offset quantization state* $k = 0, 1, 2, ...$ The values of k are indicated inside the brackets of the toryx symbols. The modified quantization equations of the imaginary e-toryces are shown in Table 14.1.

Table 14.1. Quantization equations for relative radii of spherical membranes and leading strings of excited imaginary e-toryces forming excited electrons and positrons.

Toryx	Relative radii of spherical membrane	Relative radii of leading string
$\breve{E}(k)^-_{m,n,q}$	$\breve{b}^- = 1 - 2z - k$ (14-1a)	$\breve{b}_1^- = 1 - z - k/2$ (14-1b)
$\breve{E}(k)^+_{m,n,q}$	$\breve{b}^+ = \dfrac{1}{1 - 2z - k}$ (14-2a)	$\breve{b}_1^+ = \dfrac{1 - z - k/2}{1 - 2z - k}$ (14-2b)

Notably, when $k = 0$, the equations shown in Table 14.1 reduce to the equations for excited electron and positron of basic hydrogen atom shown in Table 11.4.2.

14.2 Derivative Hydrogen Atoms – A variety of derivative hydrogen atoms is defined by the values of the toryx offset quantization state k.

Table 14.2 shows components and properties of derivative hydrogen atoms of the ordinary matter level $M = 2$ based on data from Table 13.4 and quantization equations shown in Table 14.1.

Table 14.2. Components and physical properties of derivative hydrogen atoms of the ordinary matter level $M = 2$.

k	Hydrogen atoms and their components	e_t/e	μ/μ_N	m_g/m_e	P_R
1	Nucleon core $nc_{1,1,0}^{0}$	0.000	+ 2.83120502	1921.50	+ 2.50
	Electron $e(1)_{2,1,0}^{-1}$	- 1.0	- 1835.409587	1.00	- 2.50
	Positron $e(1)_{2,1,0}^{+1}$	+ 1.0	+ 0.02444752	1.00	0.00
	Derivative hydrogen atom $^{1}H(1)_{2,1,0}^{0}$	0.000	- 1832.553934	1923.50	0.00
2	Nucleon core $nc_{1,1,0}^{0}$	0.000	+ 2.83120502	1921.50	+ 2.50
	Electron $e(2)_{2,1,0}^{-1}$	- 1.0	- 1835.415699	1.00	-3.00
	Positron $e(2)_{2,1,0}^{+1}$	+ 1.0	+ 0.02444744	1.00	0.00
	Derivative hydrogen atom $^{1}H(2)_{2,1,0}^{0}$	0.000	- 1832.560046	1923.50	- 0.50

As follows from Table 14.2, the basic and derivative hydrogen atoms have no charge. Their masses are the same and the values of their magnetic moments are either the same or almost the same. The main differences are in the values of their reality polarizations P_R.

14.3 Basic Neutron – The basic neutron $n_{1,1,0}^{0}$ of the ordinary matter level $M = 2$ is formed from the basic hydrogen atom $^{1}H_{2,1,0}^{0}$ after its excited basic electron $e_{2,1,0}^{-1}$ and excited basic positron $e_{2,1,0}^{+1}$ reduce respectively to the harmonic electron $e_{0,n,0}^{-5/4}$ and positron $e_{0,n,0}^{+5/4}$ as shown below:

$$e_{2,1,0}^{-1} \rightarrow e_{0,n,0}^{-5/4}$$
$$e_{2,1,0}^{+1} \rightarrow e_{0,n,0}^{+5/4}$$
(14.3-1)

Table 14.3. shows components and physical properties of the basic neutron $n_{1,1,0}^{0}$ of the ordinary matter level $M = 2$ based on the data from Tables 12.5, 12.6 and 13.4.

Table 14.3. Components and properties of basic neutron
of the ordinary matter level $M = 2$.

Basic neutron and its components	Symbols	μ / μ_N	m_g / m_e	P_R
Nucleon core	$nc_{1,1,0}^{0}$	+ 2.83120502	1921.50	+ 2.50
Basic harmonic electron	$e_{0,1,0}^{-5/4}$	- 14.50288888	1.25	- 2.25
Basic harmonic positron	$e_{0,1,0}^{+5/4}$	+ 4.83429629	1.25	- 0.25
Basic neutron	$n_{1,1,0}^{0}$	- 6.83738757	1924.00	± 0.00

14.4 Derivative Neutron – Derivative neutron $n(2)_{1,1,0}^{0}$ of the ordinary matter level $M = 2$ is formed from the derivative hydrogen atom $^1H(2)_{2,1,0}^{0}$ in which the excited electron $e_{2,1,0}^{-1}$ transforms into the harmonic electron $e(2)_{0,1,0}^{-1}$ as shown below:

$$e(2)_{2,1,0}^{-1} \rightarrow e(2)_{0,1,0}^{-1}$$
(14.4-1)

Table 14.4. Components and properties of derivative neutron
of the ordinary matter level $M = 2$.

Derivative neutron and its components	Symbols	μ / μ_N	m_g / m_e	P_R
Basic nucleon core	$nc_{1,1,0}^{0}$	+ 2.83120502	1921.50	+ 2.50
Derivative excited positron	$e(2)_{2,1,0}^{+1}$	+ 0.02444744	1.00	0.00
Derivative harmonic electron	$e(2)_{0,1,0}^{-1}$	- 14.13446720	1.00	- 3.00
Derivative neutron	$n(2)_{1,1,0}^{0}$	- 11.27881474	1923.50	- 0.50

Table 14.4 shows component and properties of the derivative neutron $n(2)_{1,1,0}^0$ of the ordinary matter level $M = 2$ based on data from Table 13.4 and quantization equations shown in Table 14.1.

14.5 Hydrogen Molecule – The hydrogen molecule $^2H_{m,n,q}^0$ is made up of the basic hydrogen atom $^1H_{m,n,q}^0$ and the derivative hydrogen atom $^1H(2)_{m,n,q}^0$ according to the equation:

$$^2H_{m,n,q}^0 = {}^1H_{m,n,q}^0 + {}^1H(2)_{m,n,q}^0 \qquad (14.5\text{-}1)$$

Table 14.5 shows components and properties of the hydrogen molecule $^2H_{2,1,0}^0$ of the ordinary matter level $M = 2$ based on data from Tables 13.5 and 14.2. As it follows from this table, the reality polarizations P_R of the constituent hydrogen atoms of the hydrogen molecule, the basic hydrogen atom $^1H_{2,1,0}^0$ and the derivative hydrogen atom $^1H(2)_{2,1,0}^0$, have the same magnitudes of the reality polarizations P_R but opposite signs, making the hydrogen molecule $^2H_{2,1,0}^0$ stable.

Table 14.5. Components and physical properties of derivative atoms of the ordinary matter level $M = 2$.

Hydrogen molecule and its components	Symbols	μ/μ_N	m_g/m_e	P_R
Basic hydrogen atom	$^1H_{2,1,0}^0$	- 1832.547823	1921.50	+ 0.50
Derivative hydrogen atom	$^1H(2)_{2,1,0}^0$	+ 1832.560046	1921.50	- 0.50
Hydrogen molecule	$^2H_{2,1,0}^0$	+ 0.012223	3843.00	± 0.00

Also notable is that the directions of magnetic moments of the constituent hydrogen atoms of the hydrogen molecule are opposite to one another, greatly reducing the total magnetic moment of the hydrogen molecule.

14.6 Deuterium – The deuterium $D_{2,1,0}^0$ of the ordinary matter level $M = 2$ is made up of the basic hydrogen atom $^1H_{2,1,0}^0$ and the derivative neutron $n(2)_{1,1,0}^0$ to the equation:

$$D_{2,1,0}^0 = {}^1H_{2,1,0}^0 + n(2)_{1,1,0}^0 \qquad (14.6\text{-}1)$$

Table 14.6 shows components and calculated properties of the deuterium $D^0_{2,1,0}$ of the ordinary matter level $M = 2$ based on data from Tables 13.5 and 14.4. Notably, the reality polarizations P_R of the constituents of the deuterium $D^0_{2,1,0}$ have the same magnitudes but opposite signs, making it stable.

Table 14.6. Components and properties of deuterium of the ordinary matter level $M = 2$.

Deuteriun and its components	Symbols	μ / μ_N	m_{tg} / m_e	P_R
Basic hydrogen atom	$^1H^0_{2,1,0}$	- 1832.547823	1923.5000	+ 0.50
Derivative neutron	$n(2)^0_{1,1,0}$	+ 11.27881474	1923.5000	- 0.50
Deuterium	$D^0_{2,1,0}$	**- 1821.269008**	**3847.0000**	**0.0**
Measured values:		-	3671.4858	-
Calculated/measured ratio:		-	1.0478	-

14.7 Helium 3 –

The helium 3 $He3^0_{2,1,0}$ of the matter level $M = 2$ is made up of the basic hydrogen atom $^1H^0_{2,1,0}$, the derivative hydrogen atom $^1H(2)^0_{2,1,0}$ and the basic neutron $n^0_{1,1,0}$ according to the equation:

$$He3^0_{2,1,0} = {}^1H^0_{2,1,0} + {}^1H(2)^0_{2,1,0} + n^0_{1,1,0} \qquad (14.7\text{-}1)$$

Table 14.7. Components and properties of helium 3 of the ordinary matter level $M = 2$.

Helium 3 and its components	Symbols	μ / μ_N	m_{tg} / m_e	P_R
Basic hydrogen atom	$^1H^0_{2,1,0}$	- 1832.547823	1923.5000	+ 0.50
Derivative hydrogen atom	$^1H(2)^0_{2,1,0}$	+ 1832.553934	1923.5000	- 0.50
Basic neutron	$n^0_{1,1,0}$	- 6.83738757	1924.0000	0.00
Helium 3	$He3^0_{2,1,0}$	**- 6.83127657**	**5771.0000**	**0.00**
Measured values:		-	5497.8888	-
Calculated/measured ratio:		-	1.0497	-

Table 14.7 shows components and calculated properties of the helium 3 based on data from Tables 13.5, 14.2 and 14.3. Notably, the total reality polarization P_R of the constituents of the helium 3 is equal to zero, making it stable.

14.8 <u>Stable Atoms & Isotopes</u> – Stable atoms and isotopes $X^0_{2,1,0}$ of ordinary matter level $M = 2$ are assembled in the atoms and isotopes $X^0_{2,1,0}$ according to the equation:

$$X^0_{2,1,0} = Z({}^1H^0_{2,1,0} + n(2)^0_{0,1,0}) + (A - 2Z)\, n^0_{1,1,0} \qquad (14.8\text{-}1)$$

where

     A = mass number

     Z = atomic number

     ${}^1H^0_{2,1,0}$ = basic hydrogen atom (see Table 13.5)

     $n^0_{1,1,0}$ = basic neutron (see Table 14.3)

     $n(2)^0_{1,1,0}$ = derivative neutron (see Table 14.4).

Tables 14.8.1 and 14.8.2 show the number of components in several stable light and heavy atoms and isotopes.

Table 14.8.1. Number of components in several stable light atoms and isotopes of the ordinary matter level $M = 2$.

Atoms or isotopes				Number of components		
Names	**Symbols**	Z	A	${}^1H^0_{2,1,0}$	$n(2)^0_{0,1,0}$	$n^0_{1,1,0}$
Deuterium	${}^2_1D$	1	2	1	1	0
Helium 4	${}^4_2He4$	2	4	2	2	0
Lithium 6	${}^6_3Li6$	3	6	3	3	0
Lithium 7	${}^7_3Li6$	3	7	3	3	1
Beryllium 9	${}^9_4Be9$	4	9	4	4	1
Boron 10	${}^{10}_5B10$	5	10	5	5	0
Carbon 12	${}^{12}_6C12$	6	12	6	6	0

Table 14.8.2. Number of components in several stable heavy atoms and isotopes of the ordinary matter level $M = 2$.

Atoms or isotopes				Number of components		
Names	**Symbols**	**Z**	**A**	$^1H_{2,1,0}^0$	$n(2)_{0,1,0}^0$	$n_{0,1,0}^0$
Thulium 169	$_{69}^{169}Tml169$	69	169	69	69	31
Tantalum 181	$_{73}^{181}Ta181$	73	181	73	73	35
Gold 197	$_{79}^{197}Au197$	79	197	79	79	39
Lead 208	$_{82}^{208}Pb208$	82	208	82	82	44
Bismuth 209	$_{83}^{209}Bi209$	83	209	83	83	43
Thorium 232	$_{90}^{232}Th232$	90	232	90	90	52

14.9 Oscillated neutron

Oscillated neutron $no(2)_{2,1,1}^0$ of the ordinary matter level $M = 2$ is formed from the derivative hydrogen atom $^1H(2)_{2,1,0}^0$ in which the basic excited electron $e_{2,1,0}^{-1}$ transforms into the oscillated harmonic electron $e(2)_{0,1,1}^{-1}$ as shown below:

$$e(2)_{2,1,0}^{-1} \rightarrow e(2)_{0,1,1}^{-1} \qquad (14.9\text{-}1)$$

Table 14.9. Components and properties of oscillated neutron with oscillated harmonic electron of the ordinary matter level $M = 2$.

Free neutron and its components	Symbols	μ / μ_N	m_g / m_e	P_R
Basic nucleon core	$nc_{1,1,0}^0$	+ 2.83120502	1921.5000	+ 2.50
Derivative excited positron	$e(2)_{2,1,0}^{+1}$	+ 0.02444744	1.0000	0.00
Oscillated harmonic electron	$e(2)_{0,1,1}^{-1}$	- 4.71148907	3.0000	- 9.00
Oscillated neutron	$no(2)_{2,1,1}^0$	- 1.85583661	1925.5000	- 6.50
Measured values		- 1.91304272	1838.6837	-
Calculated/measured values:		0.970097	1.0472	-

Thus, the exponential excitation quantum state m of the electron decreases from 2 to 0, while its oscillation quantum state q increases from 0 to 1.

Table 14.9 shows components and properties of the oscillated neutron $no(2)_{2,1,1}^{0}$ of the ordinary matter level $M = 2$ based on data from Table 13.4 and quantization equations shown in Table 14.1.

14.10 Proton – There are two methods to obtain the proton $p_{2,\infty,0}^{+1}$ of the ordinary matter level $M = 2$.

According to the first method, the proton $p_{2,\infty,0}^{+1}$ is formed during decay of the oscillated neutron $no(2)_{2,1,1}^{0}$. During decay of this neutron the oscillation quantum state q of its oscillated harmonic electron $e(2)_{0,n,1}^{-1}$ decreases from 1 to 0 as shown below:

$$e(2)_{0,n,1}^{-1} \rightarrow e(2)_{0,n,0}^{-1} \qquad (14.10\text{-}1)$$

The energy released during this decay is spent to expel the de-oscillated electron $e(2)_{0,n,0}^{-1}$ from the neutron, to propel the positron $e(2)_{2,1,0}^{+1}$ to the highest excitation quantum state transforming it into the excited positron $p_{2,\infty,0}^{+1}$ and also to emit electron antineutrino in the direction opposite to the direction of emitted electron.

Table 14.10. Components and properties of proton
of the ordinary matter level $M = 2$.

Free proton and its components	Symbols	μ / μ_N	m_g / m_e	P_R
Basic nucleon core	$nc_{1,1,0}^{0}$	+ 2.83120502	1921.5000	+ 2.50
Basic excited positron	$e_{2,\infty,0}^{+1}$	+ 0.00000000	1.00	± 0.00
Proton	$p_{2,\infty,0}^{+1}$	+ 2.83120502	1922.5000	+ 2.50
Measured values:		+ 2.79284736	1836.1527	-
Calculated/measured ratio:		1.0137	1.0470	-

Table 14.10 shows component and properties of the proton $p_{2,\infty,0}^{+1}$ of the ordinary matter level $M = 2$ based on data from Table 13.4 and quantization equations shown in Table 14.1.

According to the second method, the proton $p_{2,\infty,0}^{+1}$ is formed after an excited electron of either basic or derivative hydrogen atom is removed from this atom. This is equivalent to propelling the excited electron to the highest excitation quantum state. To obey the Unified Law of Stable Polarization, the excited positron of the hydrogen atom also propels to the highest excitation quantum state. The components and properties of the protons produced according to the first and second methods are the same.

CHAPTER 15

SHORT-LIVED MATTER PARTICLES

In this chapter we will describe structures and properties of two kinds of short-lived matter particles, fractrons and leptrons that do not follow the Unified Law of Stable Polarization.

15.1 Fractrons - Trons with fractional charges are called *fractrons*.

The relationship between the relative radii of leading strings b_1' and b_1'' of two constituent toryces of trons making up the fractrons with the relative fractional charge ε is defined by Eq. (11.1-1) that is rewritten below:

$$b_1'' = \frac{b_1'}{2b_1'(\varepsilon + 1) - 1}$$

The fractrons may exist at various oscillation quantum states q described by Eq. (11.6-1). According to Eq. (11.6-3), their gravitational masses increase proportionally to the toryx oscillation factor Q_q with some of its values shown in Table 11.6.1. After considering Eq. (10.2-2), we find that the relative gravitational masses of the fractron constituent toryces m_{tg}/m_e are equal to:

$$\frac{m_{tg}}{m_e} = \left| \frac{b_1 - 1}{2b_1} \right| Q_q \qquad (15.1\text{-}1)$$

A special group of fractrons with the fractional charges of $+\frac{2}{3}$ and $-\frac{1}{3}$ is called *tritrons*. The tritrons represent quarks of the Quark Model.

Table 15.1.1 shows the ranges of the relative radii of leading strings, b_1' and b_1'', of the tritron constituent toryces corresponding to their extreme (minimum or maximum) relative gravitational masses. Tables 15.1.2 - 15.1.5 show calculated gravitational masses of six tritrons Q0 – Q5 with the fractional charge $\varepsilon = +\frac{2}{3}$ and six tritrons q0 – q5 with the fractional charge $\varepsilon = -\frac{1}{3}$.

Table 15.1.1. Relative radii of leading strings of the tritron constituent toryces corresponding to the extreme values of the tritron relative gravitational masses.

Relative radii of toryx leading strings	$\varepsilon = +\frac{2}{3}$		$\varepsilon = -\frac{1}{3}$	
	Min.	**Max.**	**Min.**	**Max.**
b_1'	$\frac{1}{2}$	$+\infty$	$-\infty$	$-\frac{1}{2}$
b_1''	$\frac{3}{4}$	$\frac{3}{10}$	$\frac{3}{4}$	$\frac{3}{10}$

Table 15.1.2. Comparison of calculated ranges of masses of the tritrons Q0, Q1 and Q2 with the measured range of masses of the up quark $\left(\varepsilon = +\frac{2}{3}\right)$.

Parameters		Q0 ? quark	Q1 Up Quark	Q2 ? quark
Toryx oscillation quantum state	q	0	1	2
Toryx oscillation factor	Q_q	1.000	3.000	205.500
Calculated range of relative tritron mass	**Min.**	0.667	2.000	137.00
	Max.	1.667	5.000	342.50
Measured range of relative quark mass	Min.	?	2.935	?
	Max.	?	6.458	?

Table 15.1.3. Comparison of calculated ranges of masses of the tritrons Q3, Q4 and Q5 with the measured range of masses of the charm and top quarks $\left(\varepsilon = +\frac{2}{3}\right)$.

Parameters		Q3 Charm Quark	Q4 ? Quark	Q5 Top quark
Toryx oscillation quantum state	q	3	4	5
Toryx oscillation factor	Q_q	3519.1875	35715.236	258015.18
Calculated range of relative tritron mass	**Min.**	2346.13	23808.82	172009.5
	Max.	5865.31	59522.06	430023.6
Measured range of relative quark mass	Min.	2270.06	?	330920.5
	Max.	2622.32	?	339139.7

Table 15.1.4. Comparison of calculated ranges of masses of the tritrons q0, q1 and q2 with the measured range of masses of the down and strange quarks $(\varepsilon = -\frac{1}{3})$.

Parameters		q0 ? quark	q1 Down quark	q2 Strange quark
Toryx oscillation quantum state	q	0	1	2
Toryx oscillation factor	Q_q	1.000	3.000	205.500
Calculated range of relative tritron mass	**Min.**	**0.667**	**2.000**	**137.00**
	Max.	**2.667**	**8.000**	**548.00**
Measured range of relative quark mass	Min.	?	6.849	136.99
	Max.	?	11.742	254.40

Table 15.1.5. Comparison of calculated ranges of masses of the tritrons q3, q4 and q5 with the measured ranges of masses of the bottom quark $(\varepsilon = -\frac{1}{3})$.

Parameters		q3 Bottom quark	q4 ? quark	q5 ? quark
Toryx oscillation quantum state	q	3	4	5
Toryx oscillation factor	Q_q	3519.1875	35715.236	258015.18
Calculated range of relative tritron mass	**Min.**	**2346.13**	**23808.8**	**172009.5**
	Max.	**9384.50**	**95235.3**	**688037.8**
Measured range of relative quark mass	Min.	8082.21	?	?
	Max.	8551.88	?	?

15.2 Leptrons – The leptrons are oscillated electrons and positrons. They represent leptons of the Standard Model.

Table 15.2 compares calculated properties of negative leptrons with measured properties of respective leptons. Also shown are predicted properties of 3e- and X-trons.

Table 15.2. Comparison of calculated properties of negative leptons of the ordinary matter level $M = 2$ ($m = 2, n = 1$) with measured properties of respective leptons.

Oscillated electrons		Physical properties		
Name	Symbol	μ/μ_B	m_g/m_e	e/e_0
Electron	$e_{2,1,0}^{-1}$	**-0.99973730**	**1.000000000**	**-1.00**
Measured values:		-1.00115965	1.000000000	-1.00
Calc./meas. ratio:		0.9986	1.0000	1.00
Name	Symbol	μ/μ_{3e}	m_g/m_e	e/e_0
Negative 3e-tron	$e_{2,1,1}^{-1}$	**-0.99973730**	**3.000000000**	**-1.00**
Measured values:		?	?	-1.00
Calc./meas. ratio:		?	?	1.00
Name	Symbol	μ/μ_{μ}	m_g/m_e	e/e_0
Negative μ-tron	$e_{2,1,2}^{-1}$	**-1.00590738**	**205.5000000**	**-1.00**
Measured values		-1.00116592	206.7682841	-1.00
Calc./meas. ratio		1.0047	0.9939	1.00
Name	Symbol	μ/μ_{τ}	m_g/m_e	e/e_0
Negative τ-tron	$e_{2,1,3}^{-1}$	**-0.98788976**	**3519.187500**	**-1.00**
Measured values		?	3477.482833	-1.00
Calc./meas. ratio		?	1.0120	1.00
Name	Symbol	μ/μ_x	m_g/m_e	e/e_0
Negative x-tron	$e_{2,1,4}^{-1}$	**-0.09734680**	**3515.236100**	**-1.00**
Measured values		?	?	-1.00
Calc./meas. ratio		?	?	1.00

CHAPTER 16

BASIC CONCEPT OF HELYX

According to our theory, helyces are the prime elements of radiation particles. In this chapter we will describe the basic concept of helyx.

16.1 Helyx Basic Structure - Helyx is a double-level helicola.

    The first level of the helyx is in the form of one double-helical spiral (Fig. 16.1.1). There is an easy way to visualize the double-helical spiral. Consider two points, a and b, rotating around a pivot point m with the rotational velocity $\widetilde{V}_{1r}$, while the pivot point m propagates along a straight line O_1O_1 with the translational velocity $\widetilde{V}_{1t}$. Each point, a and b, leaves a helical spiral trace winding around the straight line O_1O_1. The two traces are separated from each other by 180^0 and form a double helix called the *helyx leading string.*

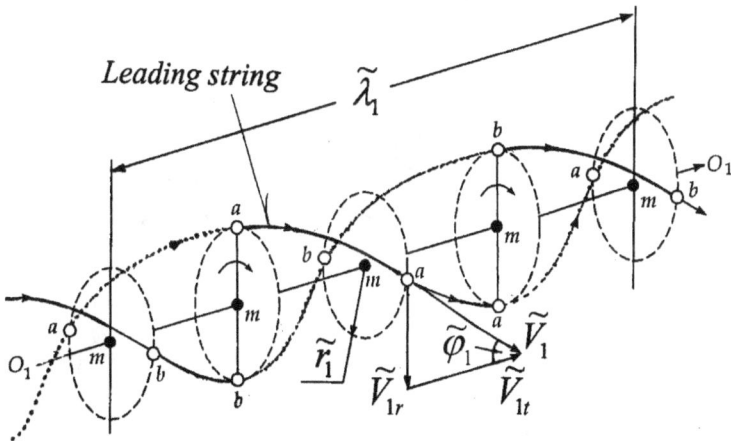

Figure 16.1.1. Structure of leading string of a helyx.

    The second level of the helyx is made up of two double helices, each wrapped around one of the helical spirals of the leading string. The two double helices of the second level of the helyx form
the *helyx trailing string.* For the sake of simplicity, Figure 16.1.2 shows a helyx with only one double-helical spiral of the leading string. There the leading string is represented by a helical trace A_1

that is accompanied by two helical traces A_2 representing the trailing string.

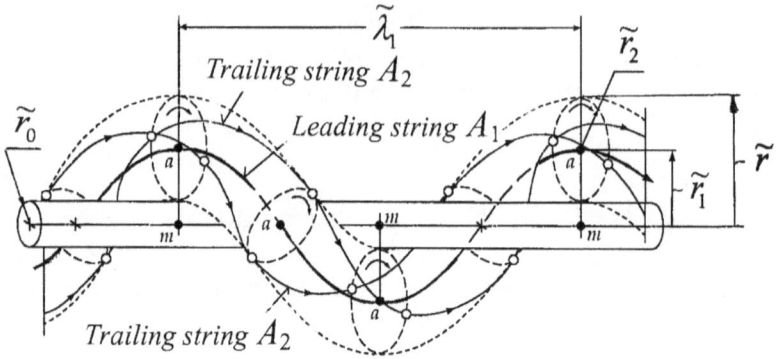

Figure 16.1.2. Structure of one of two double-helical leading strings.

16.2 Helyx Spacetime Parameters – Similarly to the toryx, the relationships between the spacetime parameters of the toryx are also described by the Pythagorean Theorem.

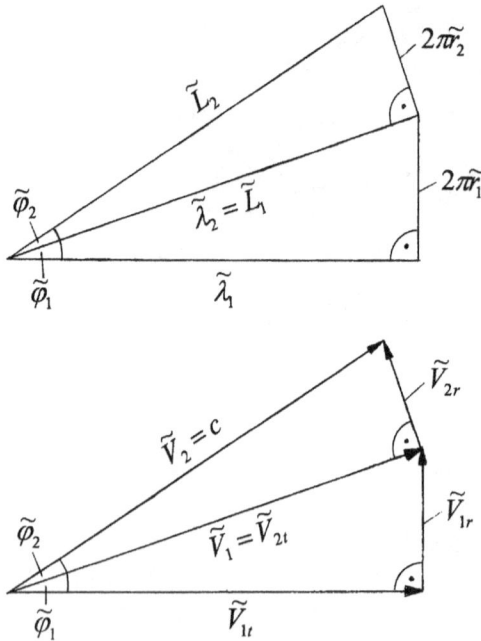

Figure 16.2. Spacetime parameters of a helyx.

In the diagrams shown in Figure 16.2 the spacetime parameters of leading and trailing strings form the sides of right triangles. Therefore, we can readily establish the relationships between these parameters by using the Pythagorean Theorem. Helical trailing strings of the helyx propagate along their helical paths in synchrony with helical leading strings, so, the translational velocity of the trailing string $\tilde{V}_{2t}$ is equal to the spiral velocity of the leading string $\tilde{V}_1$.

Notably, the symbols used for defining the helyx parameters are the same as those used for toryx, except for the "wave" mark (tilde) over the symbols of helyx parameters. Below is a list of the helyx spacetime parameters expressed in absolute values.

$\tilde{f}_0$ = helyx base frequency
$\tilde{f}_1$ = frequency of helyx leading string
$\tilde{f}_1$ = frequency of helyx trailing string
$\tilde{L}_1$ = spiral length of one winding of helyx leading string
$\tilde{L}_2$ = spiral length of one winding of helyx trailing string
$\tilde{f}_0$ = helyx peripheral radius, radius of spherical membrane
$\tilde{r}_0$ = helyx eye radius
$\tilde{r}_1$ = radius of helyx leading string
$\tilde{r}_2$ = radius of helyx trailing string
$\tilde{T}_0$ = helyx base period
$\tilde{T}_1$ = period of helyx leading string
$\tilde{T}_2$ = period of helyx trailing string
$\tilde{V}_1$ = spiral velocity of helyx leading string
$\tilde{V}_{1r}$ = rotational velocity of helyx leading string
$\tilde{V}_{1t}$ = translational velocity of helyx leading string
$\tilde{V}_2$ = spiral velocity of helyx trailing string
$\tilde{V}_{2r}$ = rotational velocity of helyx trailing string
$\tilde{V}_{2t}$ = translational velocity of helyx trailing string
$\tilde{w}_1$ = the number of windings of helyx leading string
$\tilde{w}_2$ = the number of windings of helyx trailing string
$\tilde{\lambda}_1$ = wavelength of helyx leading string
$\tilde{\lambda}_2$ = wavelength of helyx trailing string
$\tilde{\varphi}_1$ = steepness angle of helyx leading string
$\tilde{\varphi}_2$ = steepness angle of helyx trailing string.

16.3 "Helyx Sacred Code" (in absolute values) – The helyx sacred code is a set of three fundamental equations limiting the degrees of several helyx parameters, making possible to establish relationships between all spacetime parameters of the helyx.

Here is how the helyx sacred code looks when the helyx spacetime parameters are expressed in absolute values.

"Helyx Sacred Code"
(utilizing absolute values of the helyx spacetime parameters)

- The wavelength of trailing string $\tilde{\lambda}_2$ is equal to the length of one winding of leading string L_1:

$$\tilde{\lambda}_2 = \tilde{L}_1 \quad (-\infty < \tilde{r}_1 < +\infty) \qquad (16.3\text{-}1)$$

- The helyx eye radius $\tilde{r}_0$ is expressed by a positive constant value:

$$\tilde{r}_0 = \tilde{r}_1 - \tilde{r}_2 = const. \quad (-\infty < \tilde{r}_1 < +\infty) \qquad (16.3\text{-}2)$$

- The spiral velocity of the trailing string $\tilde{V}_2$ is constant and equals to the velocity of light c at each point of its spiral path. Its components, the translational velocity $\tilde{V}_{2t}$ and the rotational velocity V_{2r}, relate to the spiral velocity $\tilde{V}_2$ by the Pythagorean Theorem:

$$\tilde{V}_2 = \sqrt{\tilde{V}_{2t}^2 + \tilde{V}_{2r}^2} = c = const. \quad (-\infty < \tilde{r}_1 < +\infty) \qquad (16.3\text{-}3)$$

One can clearly see some similarity between the fundamental spacetime equations (16.3-1) – (16.3-3) for the helyx and the fundamental spacetime equations (5.3-1) – (5.3-3) for the toryx. Similarly to the toryx, it follows from equation (16.3-2) that when the radius of the helyx leading string $\tilde{r}_1$ is equal to the radius of the helyx eye $\tilde{r}_0$ the radius of the helyx trailing string $\tilde{r}_2$ reduces to zero. Consequently, the helyx transforms into a circular *real inversion helyx* with the radius $\tilde{r}_0$ and the helyx base frequency of leading string $\tilde{f}_0$.

Notably the helyx eye radius $\tilde{r}_0$ is equal to the toryx eye radius r_0, while the helyx base frequency $\tilde{f}_0$ is equal to the toryx base frequency f_0.

$$\tilde{r}_0 = r_0 \qquad (16.3\text{-}4)$$

$$\tilde{f}_0 = f_0 = \frac{c}{2\pi\tilde{r}_0} = \frac{c}{2\pi r_0} \qquad (16.3\text{-}5)$$

16.4 Derivative Spacetime Equations of Helyces - As we have done in application to the toryx, we can take advantage of the constancy of the helyx eye radius $\tilde{r}_0$, the ultimate string velocity c, and the helyx base frequency of leading string f_0 to express the helyx spacetime parameters in relative terms.

The helyx relative spacetime parameters are:

$\tilde{b} = r/r_0$ = helyx relative peripheral radius
$\tilde{b}_1 = \tilde{r}_1/\tilde{r}_0$ = relative radius of helyx leading string
$\tilde{b}_2 = \tilde{r}_2/\tilde{r}_0$ = relative radius of helyx trailing string
$\tilde{l}_1 = L_1/2\pi\tilde{r}_0$ = relative length of helyx leading string
$\tilde{l}_2 = L_2/2\pi\tilde{r}_0$ = relative length of helyx trailing string
$\tilde{t}_1 = \tilde{T}_1/\tilde{T}_0$ = relative period of helyx leading string
$\tilde{t}_2 = \tilde{T}_2/\tilde{T}_0$ = relative period of helyx trailing string
$\tilde{\beta}_1 = \tilde{V}_1/c$ = relative spiral velocity of leading string
$\tilde{\beta}_{1t} = \tilde{V}_{1t}/c$ = relative translational velocity of leading string
$\tilde{\beta}_{1r} = \tilde{V}_{1r}/c$ = relative rotational velocity of leading string
$\tilde{\beta}_2 = \tilde{V}_2/c$ = relative spiral velocity of trailing string
$\tilde{\beta}_{2t} = \tilde{V}_{2t}/c$ = relative translational velocity of trailing string
$\tilde{\beta}_{2r} = \tilde{V}_{2r}/c$ = relative rotational velocity of trailing string
$\tilde{\delta}_1 = \tilde{f}_1/\tilde{f}_0$ = relative frequency of helyx leading string
$\tilde{\delta}_2 = \tilde{f}_2/\tilde{f}_0$ = relative frequency of helyx trailing string.

With these simplifications, we can readily derive Eqs. (16.4-1) –(16.4-9) for the helyx spacetime parameters as a function of the relative radii of their leading string $\tilde{b}_1$ shown in Table 16.4. From Eq. (16.4-8) we can express the relative radius of leading string $\tilde{b}_1$ as a function of the relative frequency of trailing string $\tilde{\delta}_2$ for the two extreme cases.

Table 16.4. Spacetime parameters of helyx leading string
as a function of the relative radius of leading string $\tilde{b}_1$.

Relative parameter	Leading string Eq.(a)	Trailing string Eq. (b)
Radius Eq.(16.4-1)	$\tilde{b}_1 = \dfrac{\tilde{r}_1}{r_i}$	$\tilde{b}_2 = \tilde{b}_1 - 1$
Apex angle Eq.(16.4-2)	$\sin u(\tilde{\varphi}_1) = \dfrac{\tilde{b}_1\sqrt{2\tilde{b}_1 - 1}}{(\tilde{b}_1 - 1)^2}$	$\cos u(\tilde{\varphi}_2) = \dfrac{\tilde{b}_1 - 1}{\tilde{b}_1}$
Wave-length Eq.(16.4-3)	$\tilde{\eta}_1 = \dfrac{\tilde{\lambda}_1}{2\pi r_i} = \sqrt{\dfrac{(\tilde{b}_1 - 1)^4 - \tilde{b}_1^2(2\tilde{b}_1 - 1)}{2\tilde{b}_1 - 1}}$	$\tilde{\eta}_2 = \dfrac{\tilde{\lambda}_2}{2\pi r_i} = \dfrac{(\tilde{b}_1 - 1)^2}{\sqrt{2\tilde{b}_1 - 1}}$
Length of one winding Eq.(16.4-4)	$\tilde{l}_1 = \dfrac{\tilde{L}_1}{2\pi r_i} = \dfrac{(\tilde{b}_1 - 1)^2}{\sqrt{2\tilde{b}_1 - 1}}$	$\tilde{l}_2 = \dfrac{\tilde{L}_2}{2\pi r_i} = \dfrac{\tilde{b}_1(\tilde{b}_1 - 1)}{\sqrt{2\tilde{b}_1 - 1}}$
Transl. velocity Eq.(16.4-5)	$\tilde{\beta}_{1t} = \dfrac{\tilde{V}_{1t}}{c} = \dfrac{\sqrt{(\tilde{b}_1 - 1)^4 - \tilde{b}_1^2(2\tilde{b}_1 - 1)}}{\tilde{b}(\tilde{b}_1 - 1)}$	$\tilde{\beta}_{2t} = \dfrac{\tilde{V}_{1t}}{c} = \dfrac{\tilde{b}_1 - 1}{\tilde{b}_1}$
Rotational velocity Eq.(16.4-6)	$\tilde{\beta}_{1r} = \dfrac{\tilde{V}_{1r}}{c} = \dfrac{\sqrt{2\tilde{b}_1 - 1}}{\tilde{b}_1 - 1}$	$\tilde{\beta}_{2r} = \dfrac{\tilde{V}_{1r}}{c} = \dfrac{\sqrt{2\tilde{b}_1 - 1}}{\tilde{b}_1}$
Spiral velocity Eq.(16.4-7)	$\tilde{\beta}_1 = \dfrac{\tilde{V}_1}{c} = \dfrac{\tilde{b}_1 - 1}{\tilde{b}_1}$	$\tilde{\beta}_2 = \dfrac{\tilde{V}_2}{c} = 1$
Frequency Eq.(16.4-8)	$\tilde{\delta}_1 = \dfrac{\tilde{f}_1}{f_0} = \dfrac{\sqrt{2\tilde{b}_1 - 1}}{\tilde{b}_1(\tilde{b}_1 - 1)}$	$\tilde{\delta}_2 = \dfrac{\tilde{f}_2}{f_0} = \dfrac{\sqrt{2\tilde{b}_1 - 1}}{\tilde{b}_1(\tilde{b}_1 - 1)}$
Period Eq.(16.4-9)	$\tilde{\tau}_1 = \dfrac{\tilde{T}_1}{T_0} = \dfrac{\tilde{b}_1(\tilde{b}_1 - 1)}{\sqrt{2\tilde{b}_1 - 1}}$	$\tilde{\tau}_2 = \dfrac{\tilde{T}_2}{T_0} = \dfrac{\tilde{b}_1(\tilde{b}_1 - 1)}{\sqrt{2\tilde{b}_1 - 1}}$

$$\tilde{b}_1 = \left(\dfrac{2}{\tilde{\delta}_2^2}\right)^{1/3} \qquad (\tilde{b}_1 \gg 1) \qquad (16.4\text{-}10)$$

$$\tilde{b}_1 = -\dfrac{i}{\tilde{\delta}_2} \qquad (\tilde{b}_1 \ll 1) \qquad (16.4\text{-}11)$$

From equations shown in Table 16.4.1, we obtain:

$$\tilde{b} = 2\tilde{b}_1 - 1 \tag{16.4-12}$$

$$\tilde{l}_1 = \tilde{\eta}_2 \tag{16.4-13}$$

$$\tilde{\beta}_1 = \tilde{\beta}_{2l} \tag{16.4-14}$$

$$\tilde{\delta}_1 = \tilde{\delta}_2 \tag{16.4-15}$$

16.5 Creation of Helyces - A helyx is emitted when its "parental" toryx is transferred from a higher quantum energy state to a lower quantum energy state.

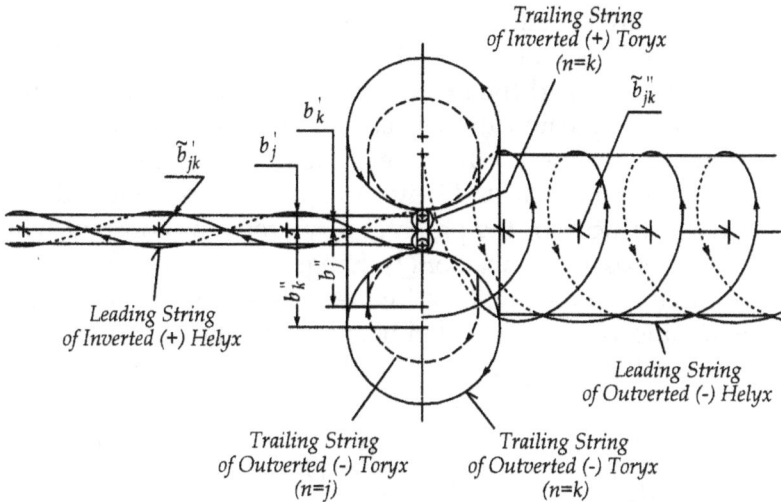

Figure 16.5. Emission of helyces.

Consider a case when quantum state of a toryx changes from a higher state $n = k$ to a lower state $n = j$ (Fig. 16.5). Let b_{1k} and b_{1j} be the respective relative radii of the toryx leading strings corresponding to these quantum states. Based on Eq. (10.4-5), the toryx exchange energies E_{xk} and E_{xj} corresponding to these quantum states are equal to:

$$E_{xj} = -m_e c^2 \frac{(b_{1j} - 1)(2b_{1j} - 1)}{4b_{1j}^3} \tag{16.5-1}$$

$$E_{xk} = -m_e c^2 \frac{(b_{1k}-1)(2b_{1k}-1)}{4b_{1k}^3} \qquad (16.5\text{-}2)$$

To obey the law of conservation of energy, the difference between the energies E_{xk} and E_{xj} must be equal to the energy of emitted helyx $\tilde{E}_{kj}$ as given by the equation:

$$\tilde{E}_{kj} = \tilde{E}_{xk} - \tilde{E}_{xj} = \frac{\tilde{f}_{1kj}h}{2} = \frac{\tilde{f}_{2kj}h}{2} \qquad (16.5\text{-}3)$$

In Eq. (16.5-3), h is the Planck constant that is equal to:

$$h = \frac{e^2 U}{2\varepsilon_0 c} \qquad (16.5\text{-}4)$$

Toryces can be either in excitation or oscillation quantum state. The helyces emitted by the excited and oscillated toryces are respectively called the *excited* and *oscillated helyces*. When an excited helyx is created the oscillation factor Q_q of the parental toryx remains constant, while the radius of its leading string b_1 changes. When an oscillated helyx is created the relative radius of the leading string b_1 of the parental toryx remains constant while its toryx toryx oscillation factor Q_q changes.

We can express the relative frequencies of leading and trailing strings $\tilde{\delta}_{1kj}$ and $\tilde{\delta}_{2kj}$ of an emitted helyx for a general case of an excited and oscillated toryx.

$$\delta_{1kj} = \delta_{2kj} = \frac{Z}{4U}\left[Q_{qk}\frac{(b_{1k}-1)(2b_{1k}-1)}{b_{1k}^3} - Q_{qj}\frac{(b_{1j}-1)(2b_{1j}-1)}{b_{1j}^3} \right] \qquad (16.5\text{-}5)$$

In application to the excited helyces in which the oscillation factor Q_q of their parental toryces remains constant, Eq. (16.5-5) reduces to the form:

$$\delta_{1kj} = \delta_{2kj} = \frac{ZQ_q}{4U}\left[\frac{(b_{1k}-1)(2b_{1k}-1)}{b_{1k}^3} - \frac{(b_{1j}-1)(2b_{1j}-1)}{b_{1j}^3} \right] \qquad (16.5\text{-}6)$$

In application to the oscillated helyces in which the relative radius of leading string b_1 of their parental toryces remains constant, Eq. (16.5-6) reduces to the form:

$$\delta_{1kj} = \delta_{2kj} = \left[\frac{Z(b_1 - 1)(2b_1 - 1)}{4Ub_1^3} \right](Q_{qk} - Q_{qj}) \qquad (16.5\text{-}7)$$

From Eqs. (16.4-8) and (16.5-3), the energy of emitted he-lyx $\tilde{E}_{kj}$ is equal to:

$$\tilde{E}_{kj} = f_i h \frac{\sqrt{2\tilde{b}_1 - 1}}{2\tilde{b}_1(\tilde{b}_1 - 1)} \qquad (16.5\text{-}8)$$

16.6 Classification of Helyces — Similarly to the toryces, we use two integrated spacetime parameters of helyces to define their main classification: the *helyx vorticity* $\tilde{V}$ and the *helyx reality* $\tilde{R}$.

Helyx vorticity - The helyx vorticity $\tilde{V}$ is equal to the ratio of the radius of trailing string $\tilde{r}_2$ to the radius of leading string $\tilde{r}_1$ with the opposite sign.

$$\tilde{V} = -\frac{\tilde{r}_2}{\tilde{r}_1} = -\frac{\tilde{b}_2}{\tilde{b}_1} = -\frac{\tilde{b}_1 - 1}{\tilde{b}_1} = -\frac{\tilde{b} - 1}{\tilde{b} + 1} = -\cos u\tilde{\varphi}_2 \qquad (16.6\text{-}1)$$

Figure 16.6.1 shows a circular diagram of the helyx vorticity $\tilde{V}$ as a function of the apex angle of trailing string $\tilde{\varphi}_2$. Toryces with positive vorticity $\tilde{V}$ are called *positive* and with negative vorticity $\tilde{V}$ *negative*.

The circular diagram of the helyx vorticity $\tilde{V}$ is divided into two domains, infinity domain and infinility domain, each occupying equal sectors of the diagram. The infinility domain occupies two top quadrants; it contains the values of $\tilde{V}$ extending clockwise from the positive unity $(+1)$ and passing through infinility (± 0) to the negative unity (-1). The infinity domain resides in two bottom quadrants; it contains the values of $\tilde{V}$ extending counterclockwise from the positive unity $(+1)$ and passing through infinity $(\pm\infty)$ to the negative unity (-1).

Notably, positive infinility $(+0)$ merges with negative infinility (-0) at $\varphi_2 \to 90^0$, while negative infinity $(-\infty)$ merges with positive infinity $(+\infty)$ at $\varphi_2 \to 270^0$. There are two kinds of symmetrical polarization between the helyx vorticities $\tilde{V}$ of the helyces that belong to the four quadrants of the circular diagram, inverse-symmetrical and reverse-symmetrical.

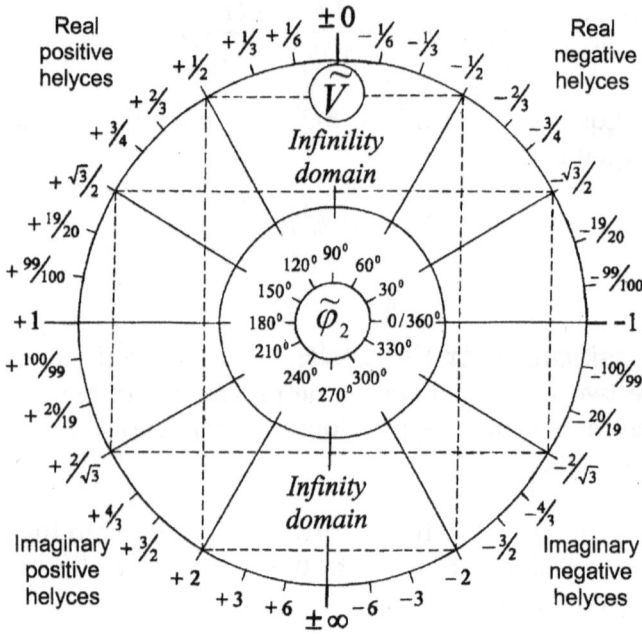

Figure 16.6.1. Helyx vorticity $\tilde{V}$ as a function of the apex angle
of trailing string $\tilde{\varphi}_2$.

The helyces that belong to the top and bottom quadrants are inverse-symmetrically polarized by their vorticity $\tilde{V}$. The magnitudes of their vorticities $\tilde{V}$ are symmetrically inversed while the signs of $\tilde{V}$ are the same. The helyces that belong to the right and left quadrants are reverse-symmetrically polarized by their vorticity $\tilde{V}$, because the signs of their vorticities $\tilde{V}$ are symmetrically reversed while the magnitudes of $\tilde{V}$ are the same.

Helyx reality - The helyx reality $\tilde{R}$ is equal to the helyx relative peripheral radius $\tilde{b}$.

$$\tilde{R} = \tilde{b} = 2\tilde{b}_1 - 1 = \frac{1 + \cos u\tilde{\varphi}_2}{1 - \cos u\tilde{\varphi}_2} \qquad (16.6\text{-}2)$$

Figure 16.6.2 shows a circular diagram of the helyx reality $\tilde{R}$ as a function of the apex angle of trailing string $\tilde{\varphi}_2$. The circular diagram of the helyx reality $\tilde{R}$ is divided into infinity and infinity domains occupying left and right quadrants respectively. There are two kinds of symmetrical polarization between the values of the helyx reality $\tilde{R}$ of the helyces that belong to the four quadrants of the circular diagram, inverse-symmetrical and reverse-symmetrical.

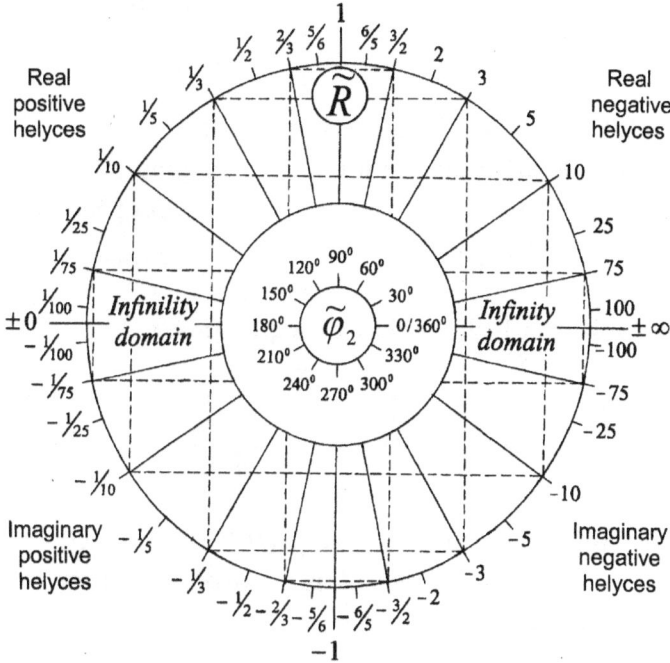

Figure 16.6.2. Helyx reality $\widetilde{R}$ as a function of the apex angle of trailing string $\widetilde{\varphi}_2$.

The helyces that belong to the right and left quadrants are inverse-symmetrically polarized by their reality $\widetilde{R}$, because the magnitudes of their realities $\widetilde{R}$ are symmetrically inversed while the signs of $\widetilde{R}$ are the same. The helyces that belong to the top and bottom quadrants are reverse-symmetrically polarized by their reality $\widetilde{R}$, because the signs of their realities $\widetilde{R}$ are symmetrically reversed while the magnitudes of $\widetilde{R}$ are the same.

Table 16.6. Main classification of helyces.

Group	Helyx	$\widetilde{\varphi}_2$	$\widetilde{V}$	$\widetilde{R}$
1	Real negative	0^0 - 90^0	(−)	(+)
2	Real positive	90^0 - 180^0	(+)	(+)
3	Imaginary positive	180^0 - 270^0	(+)	(−)
4	Imaginary negative	270^0 - 360^0	(−)	(−)

The helyces are divided into four main groups according to their vorticity $\tilde{V}$ and reality $\tilde{R}$ as shown in Table 16.6. In the real helyces, the sign of the helyx reality $\tilde{R}$ is positive, while in the imaginary helyces negative.

16.7 Helyx Golden Polarization Parameter - The helyx golden polarization parameter $\tilde{G}_h$ is equal to the product of the helyx vorticity $\tilde{V}$ and the square root of the helyx reality $\tilde{R}$ with an opposite sign.

This parameter is defined by the equation:

$$\tilde{G}_h = -\tilde{V}\sqrt{\tilde{R}} = \frac{(\tilde{b}_1 - 1)\sqrt{2\tilde{b}_1 - 1}}{\tilde{b}_1} \qquad (16.7\text{-}1)$$

Figure 16.7 shows a plot of Eq. (16.7-1) in which the helyx golden polarization parameter $\tilde{G}_h$ is expressed as a function of the relative radius of leading string $\tilde{b}_1$.

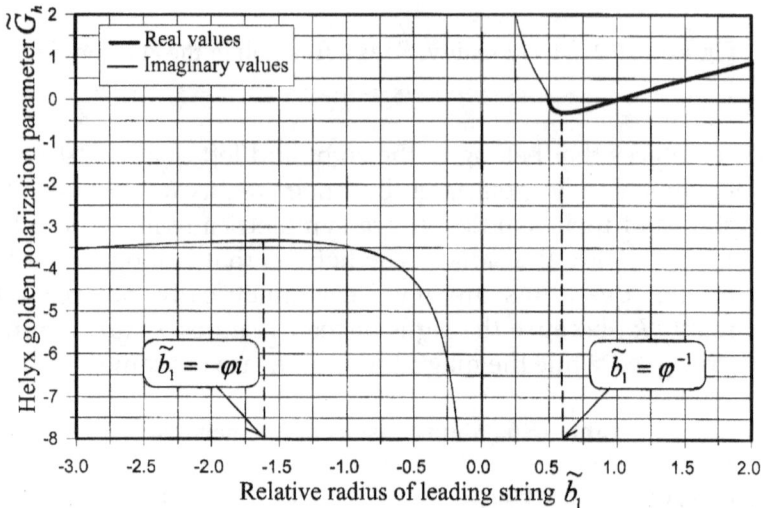

Figure 16.7. Helyx golden polarization parameter $\tilde{G}_h$.

Notably, the plot of $\tilde{G}_h$ has two extreme values of $\tilde{G}_h$ at which $\tilde{b}_1$ is directly related to the golden ratio φ. The first extreme value of $\widetilde{G}_h$ corresponds to $\widetilde{b}_1 = -\varphi i$; here the value of $\widetilde{G}_h$ is maximum. The second extreme value of $\widetilde{G}_h$ corresponds to $\widetilde{b}_1 = \varphi^{-1}$; here the value of $\widetilde{G}_l$ is minimum.

16.8 Charge and Mass of Helyx - Charge and mass of the helyx are directly related to the helyx spacetime properties that are based on three fundamental equations (16.3-1) – (16.3-3).

To express the basic physical properties of the helyx in spacetime terms we use the four additional assumptions:

- The helyx charge e_h is related to the helyx vorticity $\widetilde{V}$ by the equation:

$$\frac{e_h}{e} = -\sqrt{1-\widetilde{V}^2} = -\frac{\sqrt{2\widetilde{b}_1 - 1}}{\widetilde{b}_1} = -\sin u\widetilde{\varphi}_2 \qquad (16.8\text{-}1)$$

- The helyx gravitational mass m_{hg} is related to the helyx vorticity $\widetilde{V}$ by the equation:

$$\frac{m_{hg}}{m_e} = \left|\sqrt{1-\widetilde{V}^2}\right| = \left|\frac{\sqrt{2\widetilde{b}_1 - 1}}{\widetilde{b}_1}\right| = \left|\sin u\widetilde{\varphi}_2\right| \qquad (16.8\text{-}2)$$

- The helyx inertial mass m_{hi} is related to the helyx vorticity $\widetilde{V}$ by the equation:

$$\frac{m_{hi}}{m_e} = \sqrt{1-\widetilde{V}^2} = \frac{\sqrt{2\widetilde{b}_1 - 1}}{\widetilde{b}_1} = \sin u\widetilde{\varphi}_2 \qquad (16.8\text{-}3)$$

CHAPTER 17

TRENDS OF HELYX SPACETIME PARAMETERS

Similarly to the toryx, helyx spacetime parameters change significantly as the radius of its leading string changes from positive to negative infinity. Also similar to the toryx is the math that describes the helyx spacetime properties.

17.1 Helyx Unified Trigonometry – The trigonometry used for helyx is the same as the toryx unified trigonometry, except for one difference. In the toryx the trigonometric functions are related to the steepness angle of the toryx trailing string φ_2, while in the helyx they are related to the apex angle of the helyx trailing string $\tilde{\varphi}_2$.

Thus, for the helyx, the relationship between the helyx trigonometric function $\cos u\tilde{\varphi}_2$ and the relative radius of the helyx leading string $\tilde{b}_1$ is given by the equation:

$$\cos \tilde{\varphi}_2 = \frac{\tilde{b}_1 - 1}{\tilde{b}_1} \quad (0^0 < \tilde{\varphi}_2 < 360^0) \qquad (17.1\text{-}1)$$

Table 17.1 shows the relationship between the unified and conventional trigonometric functions in application to helyx.

Table 17.1 Relationship between unified and
conventional trigonometric functions in application to helyx.

Unified function	Conventional function	
$(0^0 < \varphi_2 < 360^0)$	$(0^0 < \varphi_2 < 180^0)$	$(180^0 < \varphi_2 < 360^0)$
$\cos u\tilde{\varphi}_2$	$\cos \varphi_2$	$\sec\varphi_2$
$\sin u\tilde{\varphi}_2$	$\sin \varphi_2$	$i\tan\varphi_2$

Below are graphical presentations of trends of the helyx spacetime parameters as a function of the apex angle of the trailing string $\tilde{\varphi}_2$.

17.2 Relative Radii of Leading and Trailing Strings –

The relative radii of leading string $\tilde{b}_1$ and trailing string $\tilde{b}_2$ are given by the equations:

$$\tilde{b}_1 = \frac{\tilde{r}_1}{r_i} = \frac{1}{1 - \cos \tilde{\varphi}_2} \qquad (0^0 < \tilde{\varphi}_2 < 360^0) \qquad (17.2\text{-}1)$$

$$b_2 = \frac{r_2}{r_i} = \frac{\cos u \tilde{\varphi}_2}{1 - \cos u \tilde{\varphi}_2} \qquad (0^0 < \tilde{\varphi}_2 < 360^0) \qquad (17.1\text{-}3)$$

Figure 17.2 shows a plot of Eqs. (17.2-1) and (17.2-3).

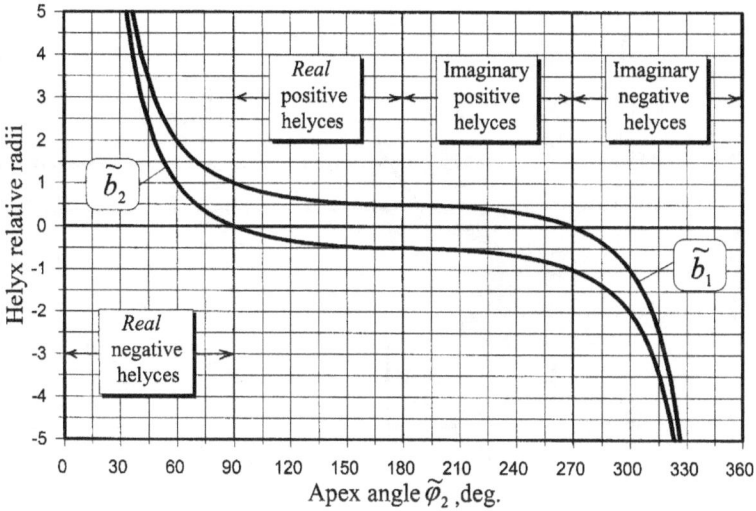

$\tilde{\varphi}_2$	$360/0^0$	90^0	180^0	270^0
$\tilde{b}_1$	$-\infty/+\infty$	$+1$	$+\frac{1}{2}$	$+0/-0$
$\tilde{b}_2$	$-\infty/+\infty$	$+0/-0$	$-\frac{1}{2}$	-1

Figure 17.2. Relative radii of leading and trailing strings, $\tilde{b}_1$ and $\tilde{b}_2$ as a function of the apex angle of trailing string $\tilde{\varphi}_2$.

17.3 Relative Wavelength of Leading String – The relative
wavelength of leading string $\tilde{\eta}_1$ is given by the equation:

$$\tilde{\eta}_1 = \frac{\tilde{\lambda}_2}{2\pi r_i} = \frac{\sqrt{\cos u^4 \tilde{\varphi}_2 - \sin u^2 \tilde{\varphi}_2}}{(1-\cos u\tilde{\varphi}_2)^2} \sqrt{\frac{1-\cos u\tilde{\varphi}_2}{1+\cos u\tilde{\varphi}_2}} \quad (0^\circ < \varphi_2 < 360^\circ) \ (17.3\text{-}1)$$

Figure 17.3 shows a plot of Eq. (17.3-1).

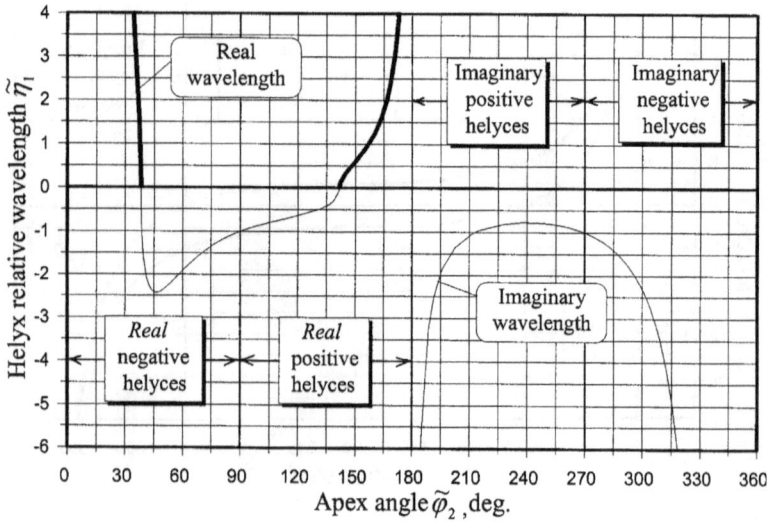

$\tilde{\varphi}_2$	$360/0^0$	90^0	180^0	270^0
$\tilde{\eta}_1$	$-\infty i/+\infty$	-1	$+\infty/-\infty i$	$-i$

Figure 17.3. Relative wavelength of leading string $\tilde{\eta}_1$ as a function of the apex angle of trailing string $\tilde{\varphi}_2$.

17.4 Relative Wavelength of Trailing String – The relative wavelength of trailing string $\tilde{\eta}_2$ is given by the equation:

$$\tilde{\eta}_2 = \frac{\tilde{\lambda}_2}{2\pi r_i} = \left(\frac{\cos u\tilde{\varphi}_2}{1 - \cos u\tilde{\varphi}_2} \right)^2 \sqrt{\frac{1 - \cos u\tilde{\varphi}_2}{1 + \cos u\tilde{\varphi}_2}} \quad (0° < \varphi_2 < 360°) \quad (17.4\text{-}1)$$

Figure 17.4 shows a plot of Eq. (17.4-1).

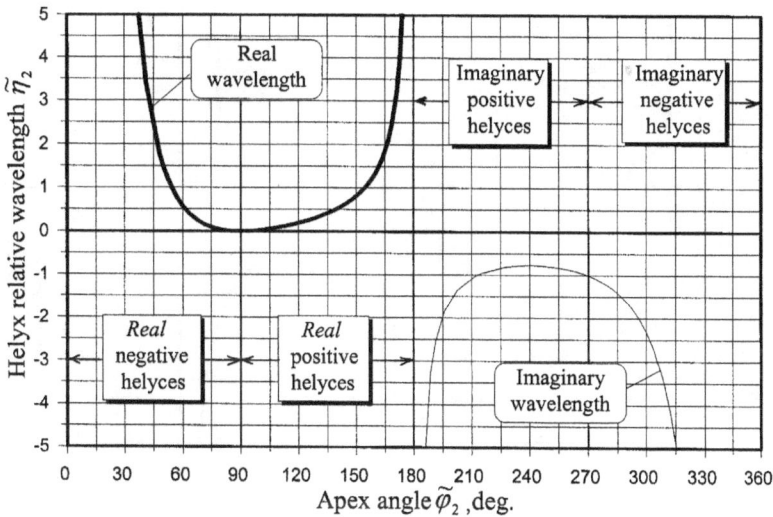

$\tilde{\varphi}_2$	360/ 0⁰	90⁰	180⁰	270⁰
$\tilde{\eta}_2$	$-\infty i / +\infty$	$+0$	$+\infty / -\infty i$	$-i$

Figure 17.4. Relative wavelength of trailing string $\tilde{\eta}_2$ as a function of the apex angle of trailing string $\tilde{\varphi}_2$.

17.5 Relative Translational Velocity of Leading String –
The relative translational velocity of leading string $\tilde{\beta}_{1t}$ is given by the equation:

$$\tilde{\beta}_{1t} = \sqrt{\cos u^2 \tilde{\varphi}_2 - \tan u^2 \tilde{\varphi}_2} \quad (0^0 < \varphi_2 < 360^0) \quad (17.5\text{-}1)$$

Figure 17.5 shows a plot of Eq. (17.5-1).

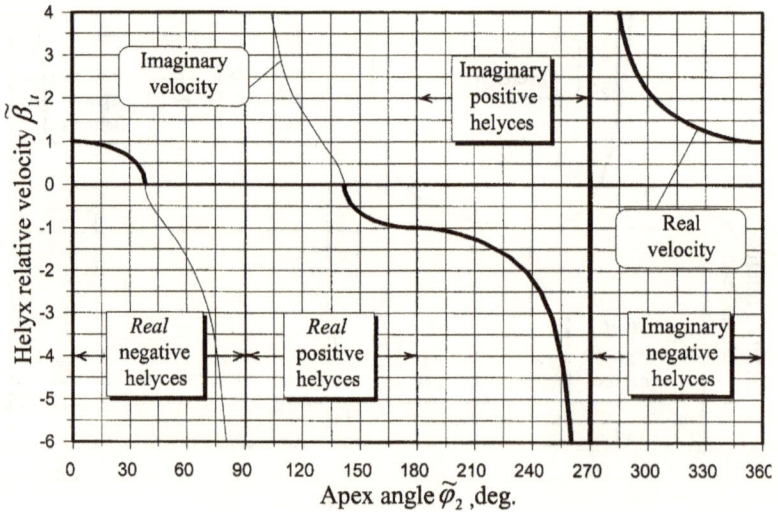

$\tilde{\varphi}_2$	$360/0^0$	90^0	180^0	270^0
$\tilde{\beta}_{1t}$	$+1$	$-\infty i / +\infty i$	-1	$-\infty / +\infty$

Figure 17.5. Relative translational velocity of leading strings $\tilde{\beta}_{1t}$ as a function of the apex angle of trailing string $\tilde{\varphi}_2$.

17.6 Relative Translational Velocity of Trailing String –

The relative translational velocity of trailing string $\tilde{\beta}_{2t}$ is given by the equation:

$$\tilde{\beta}_{2t} = \cos u\tilde{\varphi}_2 \quad (0^0 < \varphi_2 < 360^0) \tag{17.6-1}$$

Figure 17.6 shows a plot of Eqs. (17.6-1).

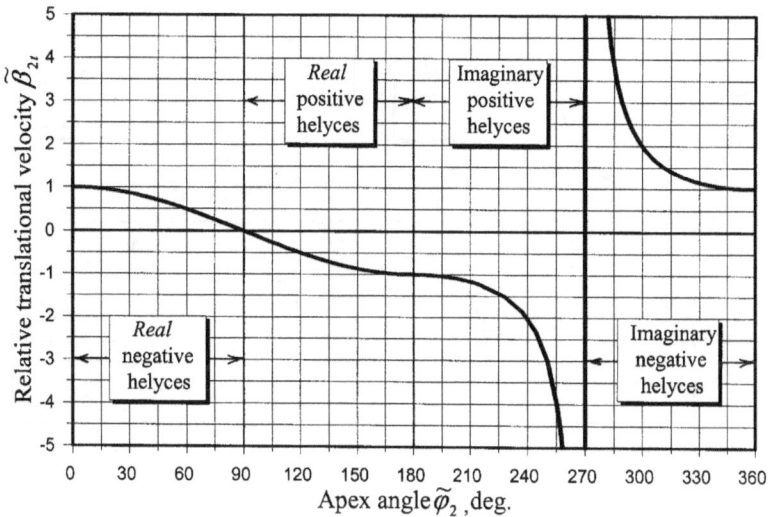

$\tilde{\varphi}_2$	$360/0^0$	90^0	180^0	270^0
$\tilde{\beta}_{2t}$	$+1$	$+0/-0$	-1	$\mp\infty$

Figure 17.6. Relative translational velocity of trailing strings $\tilde{\beta}_{2t}$ as a function of the apex angle of trailing string $\tilde{\varphi}_2$.

17.7 Relative Rotational Velocity of Leading String – The
relative rotational velocity of trailing string $\tilde{\beta}_{2r}$ is given by
the equation:

$$\tilde{\beta}_{1r} = \tan u\,\tilde{\varphi}_2 \qquad (0^0 < \varphi_2 < 360^0) \qquad (17.7\text{-}1)$$

Figure 17.7 shows a plot of Eqs. (17.7-1).

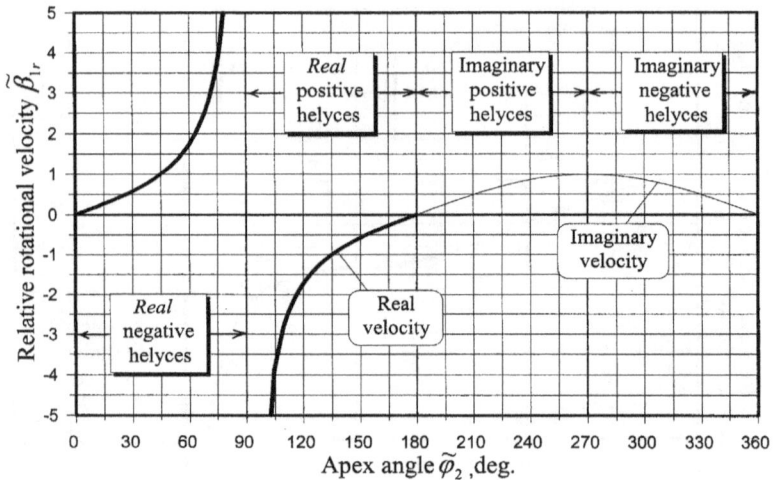

$\tilde{\varphi}_2$	$360/0^0$	90^0	180^0	270^0
$\tilde{\beta}_{1r}$	$+0i/+0$	$+\infty/-\infty$	$-0/+0i$	$+i$

Figure 17.7. Relative rotational velocity of leading strings $\tilde{\beta}_{1r}$ as
a function of the apex angle of trailing string $\tilde{\varphi}_2$.

17.8 Relative Rotational Velocity of Trailing String – The

relative rotational velocity of trailing string $\tilde{\beta}_{2r}$ is given by the equation:

$$\tilde{\beta}_{2r} = \sin u\tilde{\varphi}_2 \qquad (0^0 < \varphi_2 < 360^0) \qquad (17.8\text{-}1)$$

Figure 17.8 shows a plot of Eqs. (17.8-1).

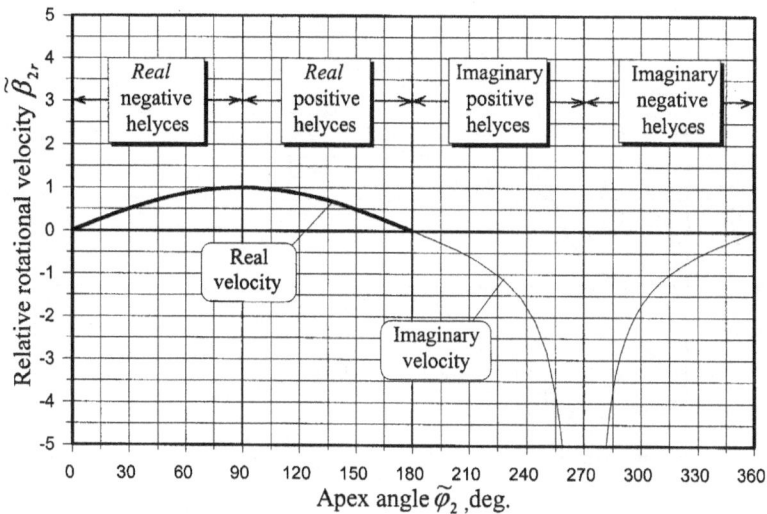

$\tilde{\varphi}_2$	$360/0^0$	90^0	180^0	270^0
$\tilde{\beta}_{2r}$	$-0i/+0$	$+1$	$+0/-0i$	$-\infty i$

Figure 17.8. Relative rotational velocity of trailing strings $\tilde{\beta}_{2r}$ as a function of the apex angle of trailing string $\tilde{\varphi}_2$.

17.9 Relative Frequency of Trailing String – The relative
frequency of trailing string $\tilde{\delta}_2$ is equal to the relative frequency of leading string $\tilde{\delta}_1$ and it is given by the equation:

$$\tilde{\delta}_2 = \tilde{\delta}_1 == \frac{(1-\cos u\tilde{\varphi}_2)^2}{\cos u\tilde{\varphi}_2}\sqrt{\frac{1+\cos u\tilde{\varphi}_2}{1-\cos u\tilde{\varphi}_2}} \quad (0^0 < \varphi_2 < 360^0) \quad (17.9\text{-}1)$$

Figure 17.9 shows a plot of Eqs. (17.9-1).

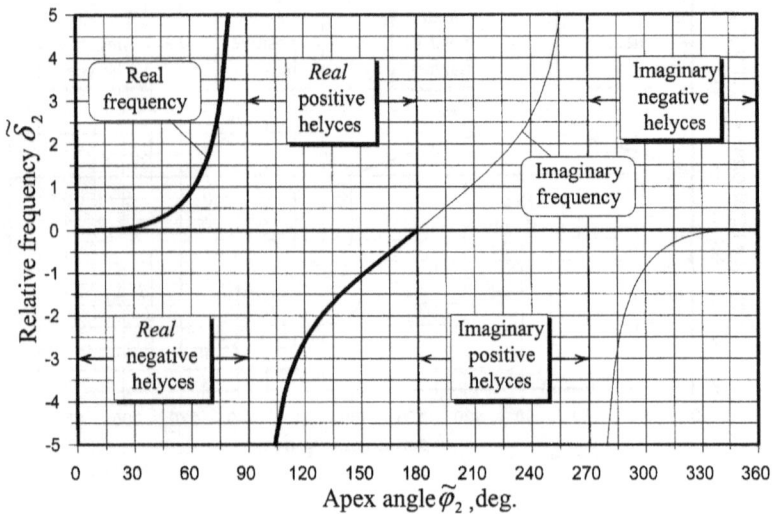

$\tilde{\varphi}_2$	$360/0^0$	90^0	180^0	270^0
$\tilde{\delta}_2$	$-0i/+0$	$+\infty/-\infty$	$-0/+0i$	$+\infty i/-\infty i$

Figure 17.9. Relative frequency of trailing strings $\tilde{\delta}_2$ as a function
of the apex angle of trailing string $\tilde{\varphi}_2$.

CHAPTER 18

INVERSION STATES OF HELYX

Similarly to the toryx, helyx spacetime topology undergoes through significant transformation as the radius of its leading string changes from positive to negative infinity. Also similar to the toryx is the math describing the helyx spacetime properties.

18.1 Metamorphoses of Helyx Topology – Topological transformations of helyces are similar to the transformations of toryces.

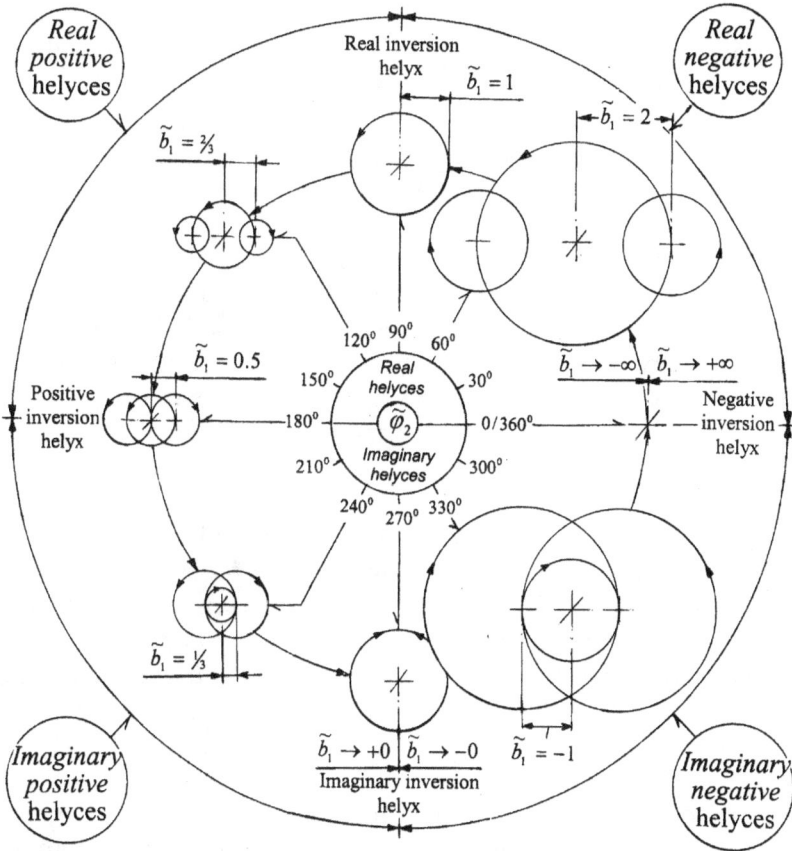

Figure 18.1. Transformations of cross-sections of helyces as a function of the apex angle of trailing string φ_2.

Figure 18.1 shows metamorphoses of cross-sections of the helyx leading and trailing strings as a function of the apex angle of
trailing string $\tilde{\varphi}_2$. Four kinds of *inversion helyces* are located at the
boundaries of the circular diagram between the four main groups
of helyces.

Negative inversion helyx $(\tilde{\varphi}_2 \to +0^\circ / 360^\circ)$ - At this point, both
leading and trailing strings become inverted. Thus, $\tilde{b} \to \pm\infty$,
$\tilde{b}_1 \to \pm\infty$ and $\tilde{b}_2 \to \pm\infty$. As $\tilde{\varphi}_2$ crosses the borderline at $0^\circ / 360^\circ$ the
helyx vorticity $\tilde{V}$ remains negative, while the helyx reality $\tilde{R}$ in-
verts from imaginary to real. The helyx appears as two parallel
lines separated by the distance equal to the diameter real inversion string.

Real inversion helyx $(\tilde{\varphi}_2 \to 90^\circ)$ - At this point, only trailing
string becomes inverted. Consequently, $\tilde{b} \to +1$, $\tilde{b}_1 \to +1$ and
$\tilde{b}_2 \to \pm0$. As $\tilde{\varphi}_2$ crosses the borderline at 90°, the helyx reality $\tilde{R}$
remains real while its vorticity $\tilde{V}$ inverts from negative to posi-
tive. It appears as circle with the relative radius $\tilde{b}_1 \to +1$.

Positive inversion helyx $(\tilde{\varphi}_2 \to 180^\circ)$ - At this point, only
spherical membrane becomes inverted. Consequently, $\tilde{b} \to \pm0$,
$\tilde{b}_1 \to +\frac{1}{2}$ and $\tilde{b}_2 \to -\frac{1}{2}$. As $\tilde{\varphi}_2$ crosses the borderline at 180°, the
helyx vorticity $\tilde{V}$ remains positive while its reality $\tilde{R}$ inverts
from real to imaginary. The helyx appears as an extreme case of a
spindle torus with inner parts of its windings touching one another.

Imaginary inversion helyx $(\tilde{\varphi}_2 \to 270^\circ)$ - At this point, only the
leading string becomes inverted. Consequently, $\tilde{b} \to -1$,
$\tilde{b}_1 \to \pm0$ and $\tilde{b}_2 \to -1$. As $\tilde{\varphi}_2$ crosses the borderline at 270°, the helyx reality $\tilde{R}$ remains imaginary while its vorticity $\tilde{V}$ inverts
from positive to negative. The helyx appears as a circle with the
relative radius approaching -1. The circle is located at the plane
perpendicular to the plane of the real inversion string

Located between the inversion helyces on the circular diagram
of Figure 18.1 are the helyces that belong to their four main
groups. Shown below are the transformations of cross-sections of
helyces within each main group.

18.2 Real Negative Helyces - Figure 18.2 shows cross-sections of the real negative helyces. They belong to the top right quadrant of the circular diagram shown in Figure 18.1.

Trailing strings of these helyces are wound counter-clockwise outside of the real inversion string. As $\widetilde{\varphi}_2$ increases, $\widetilde{b}$, $\widetilde{b}_1$ and $\widetilde{b}_2$ decrease.

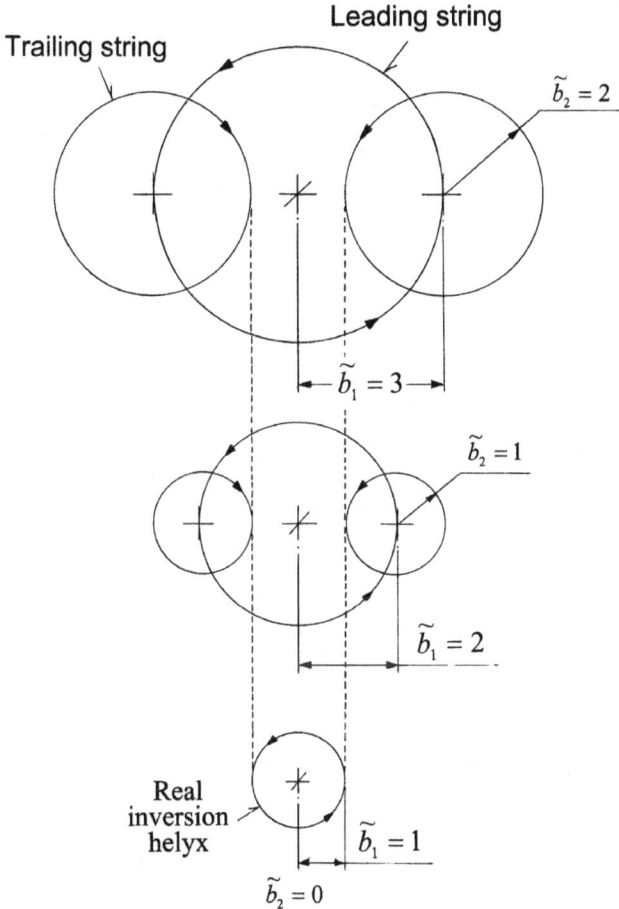

Figure 18.2. Metamorphoses of cross-sections of real negative helyces.

18.3 Real Positive Helyces - Figure 18.3 shows cross-sections of the real positive helyces. They belong to the top left quadrant of the circular diagram shown in Figure 18.1.

Within this range the trailing string is inverted, so that its windings are now wound clockwise inside the real inversion helyx. As $\tilde{\varphi}_2$ increases, both $\tilde{b}$ and $\tilde{b}_1$ decrease, but negative values of $\tilde{b}_2$ increase. Consequently, the helyx appears as an inverted helycal spiral.

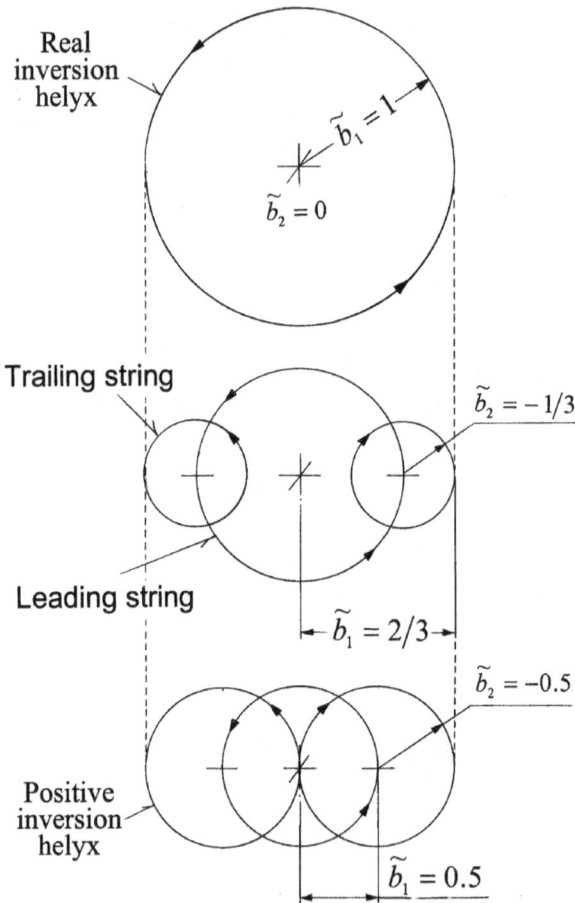

Figure 18.3. Metamorphoses of cross-sections of real positive helyces.

18.4 Imaginary Positive Helyces - Figure 18.4 shows cross-sections of the imaginary positive helyces. They belong to the bottom left quadrant of the circular diagram shown in Figure 18.1.

Here the outer helyx radius $\tilde{b}$ changes its sign from positive to negative. As $\tilde{\varphi}_2$ increases, $\tilde{b}_1$ decreases, while negative values of $\tilde{b}$ and $\tilde{b}_2$ increase. Within this range, the trailing string is still inverted and its windings are wound inside the imaginary inversion helyx. The opposite parts of windings of the trailing string intersect with one another.

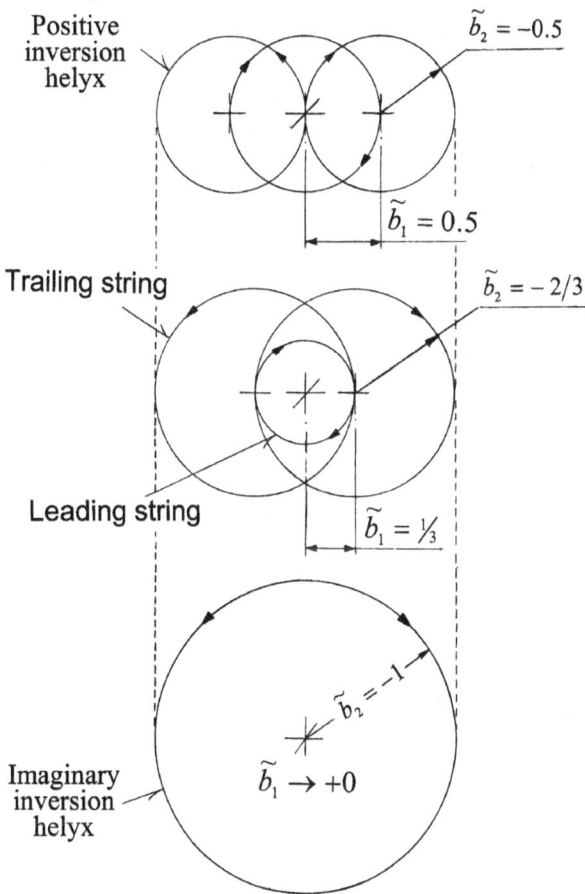

Figure 18.4. Metamorphoses of cross-sections of imaginary positive helyces.

18.5 Imaginary Negative Helyces - Figure 18.5 shows cross-sections of the imaginary negative helyces. They belong to the bottom right quadrant of the circular diagram shown in Figure 18.1.

Here the leading string becomes inverted. As the negative value of $\tilde{b}_1$ increases, the negative values of $\tilde{b}_2$ also increase. Within this range trailing string propagates outward and its windings are located outside of the imaginary inversion helyx.

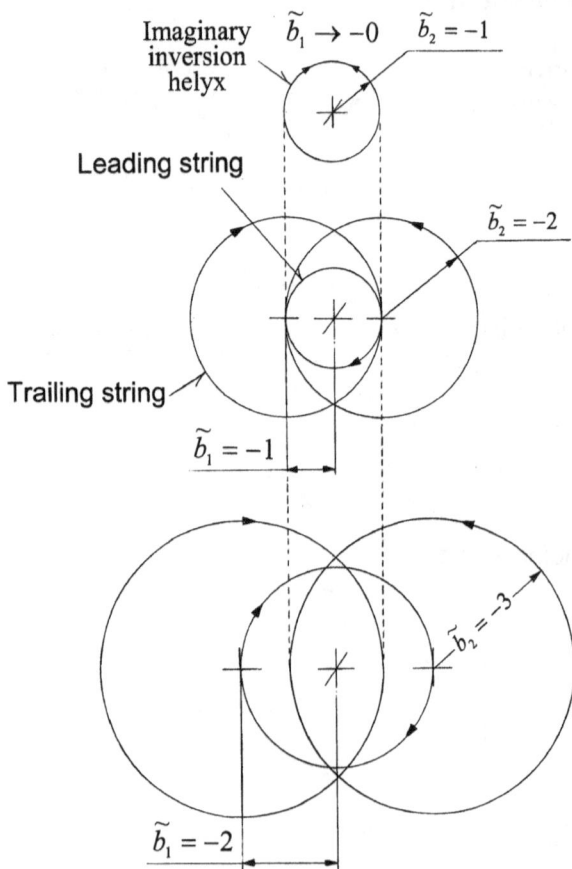

Figure 18.5. Metamorphoses of cross-sections of imaginary negative helyces.

CHAPTER 19

FORMATION OF RADIATION PARTICLES

Elementary radiation particles are composed of reality-polarized and charge-polarized matched helyces. The names of the radiation particles are similar to the names of their parental elementary particles responsible for the creation of the helyces. Elementary radiation particles emitted by excited trons are called *tons*, while the elementary radiation particles emitted by oscillated trons are called *trinos* as shown in Table 19.

Table 19. Types of elementary radiation particles.

Elementary matter particles (trons)	Elementary radiation particles	
	Excited (tons)	Oscillated (trinos)
Electron	Electon	Electrino
Positron	Positon	Positrino
Ethertron	Etherton	Ethertrino
Singulatron	Singulaton	Singulatrino

Figure 19 shows schematically the formation of four elementary radiation particles (tons) from polarized helyces.

Electons are the reality-polarized negative electons $^{-}\tilde{e}_{m,n,q}^{m,n,q}$. They are made up of the real negative helyces $^{-}\tilde{E}_{m,n,q}^{m,n,q}$ and the imaginary negative helyces $^{-}\tilde{E}_{m,n,q}^{m,n,q}$.

Positons are the reality-polarized positive electons $^{+}\tilde{e}_{m,n,q}^{m,n,q}$. They are made up of the real positive helyces $^{+}\tilde{E}_{m,n,q}^{m,n,q}$ and the imaginary positive helyces $^{+}\tilde{E}_{m,n,q}^{m,n,q}$.

Ethertons are the charge-polarized real ethertrons $^{0}\tilde{a}_{m,n,q}^{m,n,q}$. They are made up of the real negative helyces $^{-}\tilde{A}_{m,n,q}^{m,n,q}$ and the real positive helyces $^{+}\tilde{A}_{m,n,q}^{m,n,q}$.

Singulatons are the charge-polarized imaginary ethertrons $^{0}\tilde{a}_{m,n,q}^{m,n,q}$. They are made up of the imaginary negative helyces $^{-}\tilde{A}_{m,n,q}^{m,n,q}$ and the imaginary positive helyces $^{+}\tilde{A}_{m,n,q}^{m,n,q}$.

In the symbols of helyces and tons, the right superscripts and subscripts indicate respectively the values of the quantum states m, n and q of their parental toryces and trons prior to and after their emission of helyces and tons. The left superscripts indicate the sign of charges of helyces and tons.

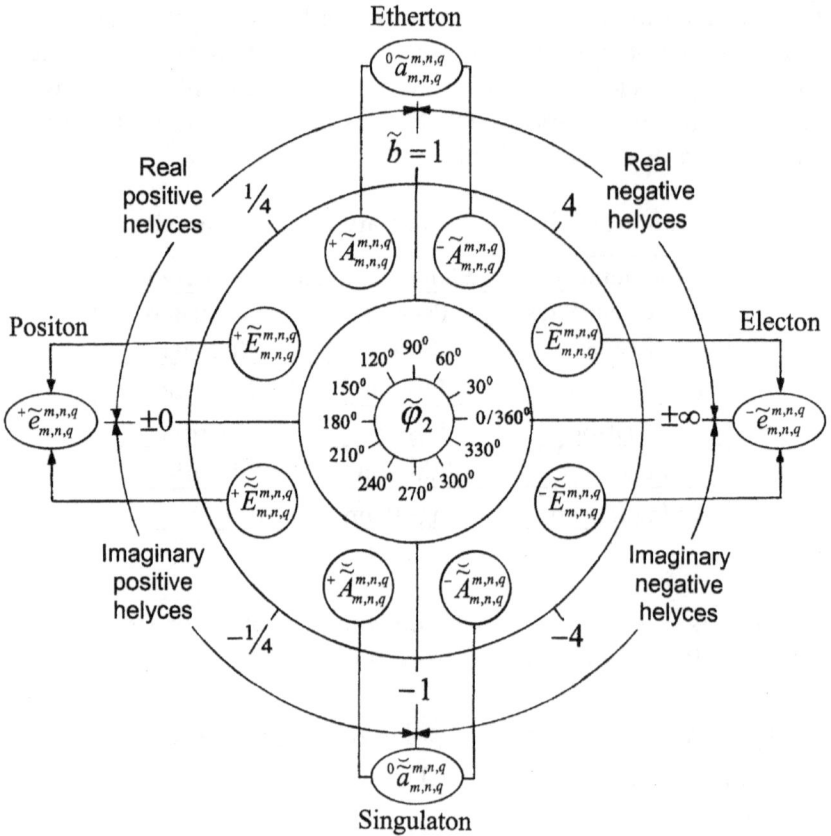

Figure 19. Formation of four elementary radiation particles (tons) from polarized helyces.

19.1 Excited Electons & Positons –
Excited electons and positons are emitted when their respective parental excited electrons and positrons are transferred from the higher quantum states n_k to the lower quantum states n_j.

<u>Excited electons</u> - As we described in Chapter 12, excited electrons are composed of negative reality-polarized matched toryces. Quantization equations (11.3-1), (11.3-6b) - (11.3-7b) describe exci-

tation quantum values of relative radii of their toryx leading strings b_1.

Table 19.1.1 shows the parameters of the matched toryces that compose the excited electrons ($k = 0$) of ordinary matter ($M = 2$) in the linear excitation quantum states n. Among the shown toryx parameters are the relative rotational velocities β_{2r} and the frequencies f_2 of the real and imaginary trailing strings. In the real toryx of an electron, the rotational velocity β_{2r} is slightly less than the velocity of light ($\beta_{2r} < 1$), while in the imaginary toryx, β_{2r} is slightly greater than the velocity of light ($\beta_{2r} > 1$).

Table 19.1.1. Compositions and properties of excited electrons ($k = 0$) of ordinary matter ($M = 2$) in the linear excitation quantum states n.

Electrons		Toryces of excited electrons			
n	Symbol	Symbol	b_1	β_{2r}	f_2, Hz
1	$e(0)_{2,1,0}^{-1}$	$E(0)_{2,1,0}^{-}$	37538.0	0.99997336	9.0213×10^{17}
		$\breve{E}(0)_{2,1,0}^{-}$	-37558.0	1.00002664	-9.0213×10^{17}
2	$e(0)_{2,2,0}^{-1}$	$E(0)_{2,2,0}^{-}$	150152.0	0.99999334	2.2553×10^{17}
		$\breve{E}(0)_{2,2,0}^{-}$	-150152.0	1.00000666	-2.2553×10^{17}
3	$e(0)_{2,3,0}^{-1}$	$E(0)_{2,3,0}^{-}$	337842.0	0.99999704	1.0024×10^{17}
		$\breve{E}(0)_{2,3,0}^{-}$	-338042.0	1.00000296	-1.0024×10^{17}
4	$e(0)_{2,4,0}^{-1}$	$E(0)_{2,4,0}^{-}$	600608.0	0.99999834	5.6383×10^{16}
		$\breve{E}(0)_{2,4,0}^{-}$	-600607.8	1.00000167	-5.6383×10^{16}

As shown in Figure 19.1, the real toryx R of excited electron emits a real subluminal helyx in which the translational velocity of trailing string $\tilde{\beta}_{2t}$ is subluminal. At the same time, an imaginary toryx I emits in the same direction an imaginary superluminal helyx in which $\tilde{\beta}_{2t}$ is superluminal. The real and imaginary helyces have opposite spins, and a reality-polarized electon formed by these helyces propagates with the average velocity equal to the velocity of light c.

Excited electron   Real subluminal helyx

R

Toryces

I

Electon

Imaginary superluminal helyx

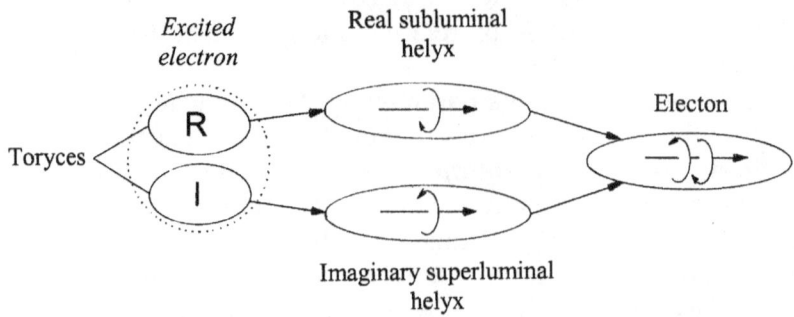

Figure 19.1. Creation of an electon.

Table 19.1.2. Compositions and properties of electons ($k = 0$) emitted by excited electrons of various matter levels M.

Electons Symbols	Helyces of excited electrons			
	Symbols	$\widetilde{b}_1$	$\widetilde{\beta}_{2i}$	$\widetilde{f}_1$, Hz
$-\widetilde{e}(0)_{1,1,0}^{1,2,0}$	$-\widetilde{E}(0)_{1,1,0}^{1,2,0}$	3580.014	0.999720672	2.236×10^{17}
$M = 1$	$-\widetilde{E}(0)_{1,1,0}^{1,2,0}$	-3540.036	1.000282483	-2.273×10^{17}
$-\widetilde{e}(0)_{1,2,0}^{1,3,0}$	$-\widetilde{E}(0)_{1,2,0}^{1,3,0}$	7428.020	0.999865375	7.482×10^{16}
$M = 1$	$-\widetilde{E}(0)_{1,2,0}^{1,3,0}$	-7381.977	1.000135465	-7.550×10^{16}
$-\widetilde{e}(0)_{2,1,0}^{2,2,0}$	$-\widetilde{E}(0)_{2,1,0}^{2,2,0}$	72201.102	0.999986150	2.469×10^{15}
$M = 2$	$-\widetilde{E}(0)_{2,1,0}^{2,2,0}$	-72195.293	1.000013851	-2.469×10^{15}
$-\widetilde{e}(0)_{2,2,0}^{2,3,0}$	$-\widetilde{E}(0)_{2,2,0}^{2,3,0}$	222225.563	0.999995500	4.572×10^{14}
$M = 2$	$-\widetilde{E}(0)_{2,2,0}^{2,3,0}$	-222220.288	1.000004500	-4.572×10^{14}
$-\widetilde{e}(0)_{3,1,0}^{3,2,0}$	$-\widetilde{E}(0)_{3,1,0}^{3,2,0}$	1.731×10^6	0.999999422	2.102×10^{13}
$M = 3$	$-\widetilde{E}(0)_{3,1,0}^{3,2,0}$	-1.731×10^6	1.000000578	-2.102×10^{13}
$-\widetilde{e}(0)_{3,2,0}^{3,3,0}$	$-\widetilde{E}(0)_{3,2,0}^{3,3,0}$	8.008×10^6	0.999999875	2.113×10^{12}
$M = 3$	$-\widetilde{E}(0)_{3,2,0}^{3,3,0}$	-8.008×10^6	1.000000125	-2.113×10^{12}

Table 19.1.2 shows compositions and properties of electons ($k = 0$) emitted by excited electrons in the exponential excitation quantum states m and the linear excitation quantum states n.

Table 19.1.3. Regions of the electromagnetic radiation.

Region	Wavelength cm	Frequency Hz
Radio waves	> 10	$< 3 \times 10^9$
Microwave	$10 - 10^{-2}$	$3 \times 10^9 - 3 \times 10^{12}$
Infrared	$10^{-2} - 7 \times 10^{-5}$	$3 \times 10^{12} - 4.3 \times 10^{14}$
Visible	$7 \times 10^{-5} - 4 \times 10^{-5}$	$4.3 \times 10^{14} - 7.5 \times 10^{14}$
Ultraviolet	$4 \times 10^{-5} - 10^{-7}$	$7.5 \times 10^{14} - 3 \times 10^{17}$
X-rays	$10^{-6} - 10^{-9}$	$3 \times 10^{16} - 3 \times 10^{21}$
Gamma rays	$< 10^{-9}$	$3 \times 10^{18} - 3.4 \times 10^{22}$

Based on the data shown in Tables 19.1.2 and 19.1.3 we can derive several important conclusions regarding to the properties of the electons corresponding to the exponential excitation quantum states m and the matter levels M.

- The relative radii of the leading strings $\tilde{b}_1$ of the helyces making up the electons increase with the increase of the exponential quantum state m.
- The frequencies of the electons $\tilde{f}_{kj}$ decrease with the increase of m.
- For $M = 1$, the calculated frequencies of the electons are within the frequency range of X-rays and ultraviolet rays.
- For $M = 2$, the calculated frequencies of the electons are very close to the measured frequencies emitted by atomic electrons of the hydrogen atom within visible and infrared frequency ranges.
- For $M = 3$, the calculated frequencies of the electons are within infrared and microwave frequency ranges.
- For $M = 4$, calculated frequencies of the electons are within microwave and radio frequency ranges.

Let us compare the calculated frequencies of electons emitted by atomic electrons of hydrogen atom with the measured frequencies of the spectra lines for hydrogen atom. The frequencies $\tilde{f}_{kj}$ of some spectra lines for hydrogen atom are accurately described by Rydberg's equation proposed by the Swiss spectroscopist Johannes Robert Rydberg:

$$\tilde{f}_{kj} = R_\infty c \left(\frac{1}{n_j^2} - \frac{1}{n_k^2} \right) \qquad (19.1\text{-}1)$$

where R_∞ is the Rydberg constant.

Table 19.1.4 shows a comparison of the calculated frequencies $\tilde{f}_{kj}$ of the spectra lines for hydrogen atom with those obtained from Rydberg's equation (19.1-1).

Table 19.1.4. Comparison of calculated frequencies of spectra lines for hydrogen atom with those obtained from Rydberg's equation (19.2-1).

Quantum states		Spectra lines of hydrogen atom	Frequencies $\tilde{f}_{kj}$, Hz		Calculated/ measured ratio
n_k	n_j	Symbols	Calculated	Measured	
3	2	H_α	4.5717×10^{14}	4.5668×10^{14}	1.0005
4	2	H_β	6.1717×10^{14}	6.1651×10^{14}	1.0005
5	2	H_γ	6.9123×10^{14}	6.9049×10^{14}	1.0005
6	2	H_δ	7.3146×10^{14}	7.3068×10^{14}	1.0005

Excited positons - As we described in Chapter 12, excited positrons are composed of negative reality-polarized matched toryces. Quantization equations (11.3-1), (11.3-8b) - (11.3-9b) describe excitation quantum values of relative radii of their toryx leading strings b_1. When positrons are transferred from a higher quantum state n_k to the lower quantum state n_j, they emit positons with the following properties:

- Magnitudes of frequencies of emitted positons are the same as the electons emitted by the electrons transferred between the same quantum states, but the signs of the frequencies are reversed.
- The relative radii of leading strings $\tilde{b}_1$ of real helyces making up positons are slightly greater than 0.5, while for the helyces making up imaginary helyces of the positons $\tilde{b}_1$ is slightly less than 0.5.

- The relative translational velocities of leading and trailing strings of real helyces making up positons are slightly less than 1.0, while for the helyces making up imaginary helyces of the positons these velocities are slightly less than 1.0.

19.2 Electrinos & Positrinos – Electrinos and positrinos are emitted when their respective parental oscillated electrons and positrons are transferred from the higher oscillation quantum states q_k to the lower oscillation quantum states q_j.

The oscillated electrons and positrons are made up of real and imaginary matched toryces. The toryx oscillation factor Q_q in various oscillation quantum states q is defined by Eq. (11.6-1).

Table 19.2.1. Compositions and properties of oscillated electrons of the ordinary matter ($M = 2, k = 0, m = 2, n = 1$).

Electrons		Toryces of oscillated electrons			
Q	Name	Symbols	b_{10}	β_{2r}	F_2, Hz
0	**Electron** $e(0)_{1,1,0}^{-1}$	$E(0)_{1,1,0}^-$	37538.000	0.999973360	9.0213×10^{17}
		$\breve{E}(0)_{1,1,0}^-$	-37538.000	1.000026640	-9.0213×10^{17}
1	**3e-tron** $e(0)_{1,1,1}^{-1}$	$E(0)_{1,1,1}^-$	12512.667	0.999973360	2.7064×10^{18}
		$\breve{E}(0)_{1,1,1}^-$	-12512.667	1.000026640	-2.7064×10^{18}
2	**μ-tron** $e(0)_{1,1,2}^{-1}$	$E(0)_{1,1,2}^-$	182.66667	0.999973360	1.8539×10^{20}
		$\breve{E}(0)_{1,1,2}^-$	-182.66667	1.000026640	-1.8539×10^{20}
3	**τ-tron** $e(0)_{1,1,3}^{-1}$	$E(0)_{1,1,3}^-$	10.66667	0.999973360	3.1748×10^{21}
		$\breve{E}(0)_{1,1,3}^-$	-10.66667	1.000026640	-3.1748×10^{21}
4	**x-tron** $e(0)_{1,1,4}^{-1}$	$E(0)_{1,1,4}^-$	1.051095	0.999973360	3.2218×10^{22}
		$\breve{E}(0)_{1,1,4}^-$	-1.051095	1.000026640	-3.2218×10^{22}

Table 19.2.1 shows compositions and properties of oscillated electrons ($M = 2, k = 0, m = 2, n = 1$) in various oscillation quantum states q. The relative radii of the toryx leading strings b_{10} are expressed in respect to the radius of real inversion toryx r_{i0} corresponding to $q = 0$. Among the other toryx parameters shown in Table 19.2.1 are the relative rotational velocities β_{2r} and the frequencies f_2 of the real and imaginary trailing strings. In the real

toryces of electrons, the rotational velocity β_{2r} is slightly less than the velocity of light ($\beta_{2r} < 1$), while in the imaginary toryces β_{2r} is slightly greater than the velocity of light ($\beta_{2r} > 1$).

Table 19.2.2 shows compositions and properties of electrinos emitted by oscillated electrons of the ordinary matter ($M = 2$, $m = 2$, $n = 1$). When toryces of oscillated electrons and positrons are transferred from a higher oscillation quantum state q_k to the lower oscillation quantum state q_j, they emit in the same direction real subluminal helyces and superluminal imaginary helyces. Consequently, the created respective electrinos and positrinos propagate with the average velocity equal to the velocity of light.

Table 19.2.2. Compositions and properties of electrinos emitted by oscillated electrons of the ordinary matter ($M = 2$, $m = 1$, $n = 1$).

Electrinos	Helyces of electrinos			
Symbols	Symbols	$\widetilde{b}_1$	$\widetilde{\beta}_{2t}$	$\widetilde{f}_1$, Hz
$-\widetilde{e}_{2,1,0}^{2,1,1}$	$-\widetilde{E}_{2,1,0}^{2,1,1}$	37546.0758	0.99997336	-6.5829×10^{15}
	$\widetilde{E}_{2,1,0}^{2,1,1}$	-37543.0754	1.00002664	6.5834×10^{15}
$-\widetilde{e}_{2,1,1}^{2,1,2}$	$-\widetilde{E}_{2,1,1}^{2,1,2}$	1728.8383	0.99942158	-6.6652×10^{17}
	$-\widetilde{E}_{2,1,1}^{2,1,2}$	-1727.7462	1.00057879	6.6657×10^{17}
$-\widetilde{e}_{2,1,2}^{2,1,3}$	$-\widetilde{E}_{2,1,2}^{2,1,3}$	268.64584	0.99627763	-1.0907×10^{19}
	$-\widetilde{E}_{2,1,2}^{2,1,3}$	-267.63156	1.00373648	1.0908×10^{19}
$-\widetilde{e}_{2,1,2}^{2,1,3}$	$-\widetilde{E}_{2,1,2}^{2,1,3}$	59.39273	0.98316392	-1.0597×10^{20}
	$\widetilde{E}_{2,1,2}^{2,1,3}$	-58.38959	1.01712634	1.0597×10^{20}

19.3 Excited Ethertons –

The excited ethertons are emitted when excited parental ethertrons are transferred from the higher quantum states n_k to the lower quantum states n_j.

As we described in Chapter 12, the excited ethertrons are made up of charge-polarized matched toryces. The relative radii of the toryx leading strings b_1 are defined by the quantization equations (11.3-1), (11.3-2b) - (11.3-3b)

Table 19.3.1 shows compositions and properties of ethertrons of the ordinary matter $M = 2$ ($m = 1$) in several linear excitation quantum states n. Among the shown parameters are the relative rotational velocities β_{2r} and the frequencies f_2, of positive and negative trailing strings. The rotational velocities of these strings have opposite signs; these velocities are very small in comparison with the velocity of light ($\beta_{2r} \ll 1$), while the translational velocities are very close to the velocity of light.

Table 19.3.1. Compositions and properties of excited ethertrons of the ordinary matter level $M = 2$ ($m = 1$) in excitation quantum states n.

Ethertrons		Toryces of excited ethertrons			
N	Symbol	Symbol	b_1	β_{2r}	f_2, Hz
1	$a_{2,1,0}^0$	$A_{2,1,0}^-$	1.00366300	0.003649635	3.3740×10^{22}
		$A_{2,1,0}^+$	0.99636363	-0.003649635	3.3988×10^{22}
2	$a_{2,2,0}^0$	$A_{2,2,0}^-$	1.00182815	0.001824818	3.3802×10^{22}
		$A_{2,3,0}^+$	0.99817851	-0.001824818	3.3926×10^{22}
3	$a_{2,3,0}^0$	$A_{2,3,0}^-$	1.00121803	0.001216545	3.3823×10^{22}
		$A_{2,3,0}^+$	0.99878493	- 0.001216545	3.3905×10^{22}
4	$a_{2,4,0}^0$	$A_{2,4,0}^-$	1.00091324	0.000912409	3.3833×10^{22}
		$A_{2,4,0}^+$	0.99908842	-0.000912409	3.3895×10^{22}

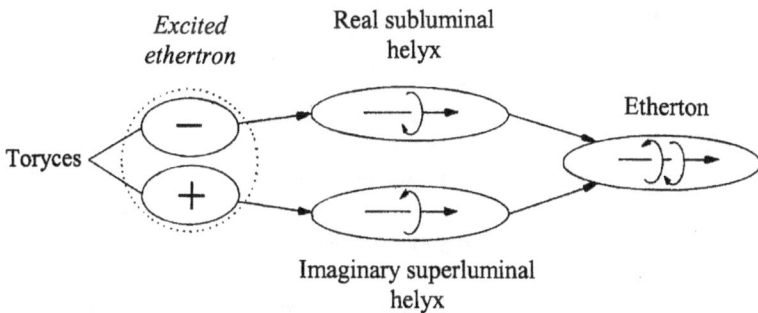

Figure 19.3. Creation of etherton.

As shown in Figure 19.3, a negative real toryx of excited real ethertron emits a real helyx in which the translational velocity of trailing string $\tilde{\beta}_{2t}$ is subluminal. At the same time, a positive real toryx emits in the same direction an imaginary helyx in which $\tilde{\beta}_{2t}$ is superluminal. The real and imaginary helyces have opposite spins, and the reality-polarized etherton formed by these helyces propagates at the average velocity equal to the velocity of light c. Ethertrons emit two kinds of ethertons, one with large radii of leading strings and the other one with small radii, both propagating at the same frequency.

Table 19.3.2. Compositions and properties of large-radius ethertons emitted by excited ethertrons of the ordinary mater level $M = 2$ ($m = 1$).

Ethertons	Large-radius etherton helyces			
Symbols	Symbols	$\tilde{b}_1$	$\tilde{\beta}_{2t}$	$\tilde{f}_1$, Hz
$^0\tilde{a}^{2,2,0}_{2,1,0}$	$-\tilde{A}^{2,2,0}_{2,1,0}$	5651.6074	0.99982306	1.1273×10^{17}
	$+\tilde{A}^{2,2,0}_{2,1,0}$	-5650.6074	1.00017697	-1.1273×10^{17}
$^0\tilde{a}^{2,3,0}_{2,2,0}$	$-\tilde{A}^{2,3,0}_{2,2,0}$	11755.1496	0.99991493	3.7579×10^{16}
	$+\tilde{A}^{2,3,0}_{2,2,0}$	-11754.1496	1.00008508	-3.7579×10^{16}
$^0\tilde{a}^{2,4,0}_{2,3,0}$	$-\tilde{A}^{2,4,0}_{2,3,0}$	18659.7982	0.99994641	1.8789×10^{16}
	$+\tilde{A}^{2,4,0}_{2,3,0}$	-18658.7982	1.00005359	-1.8789×10^{16}

Table 19.3.3. Compositions and properties of small-radius ethertons emitted by excited ethertrons of the ordinary mater level $M = 2$ ($m = 1$).

Ethertons	Small-radius etherton helyces	
Symbols	Symbols	$\tilde{b}_1$	$\tilde{\beta}_{2t}$
$^0\tilde{a}^{2,2,0}_{2,1,0}$	$-\tilde{A}^{2,2,0}_{2,1,0}$	0.500000000000346	-0.99999999999862
	$+\tilde{A}^{2,2,0}_{2,1,0}$	0.499999999999654	-1.00000000000138
$^0\tilde{a}^{2,3,0}_{2,2,0}$	$-\tilde{A}^{2,3,0}_{2,2,0}$	0.500000000000038	-0.99999999999985
	$+\tilde{A}^{2,3,0}_{2,2,0}$	0.499999999999962	-1.00000000000015
$^0\tilde{a}^{2,4,0}_{23,0}$	$-\tilde{A}^{2,4,0}_{2,3,0}$	0.500000000000008	-0.99999999999997
	$+\tilde{A}^{2,4,0}_{2,3,0}$	0.499999999999992	-1.00000000000003

Table 19.3.2 shows compositions and properties of large-radius etherton helyces emitted by excited ethertrons of the ordinary matter level $M = 2$ ($m = 1$). Table 19.3.3 shows compositions and properties of small-radius etherton helyces emitted by excited ethertrons of the ordinary matter level $M = 2$ ($m = 1$).

19.4 Excited Singulatons – The excited singulatons are emitted when excited singulatrons are transferred from the higher quantum states n_k to the lower quantum states n_j .

As we described in Chapter 12, the singulatrons are made up of imaginary charge-polarized matched toryces. The relative radii of the toryx leading strings b_1 in various excitation quantum states are defined by the quantization (11.3-1), (11.3-4b) - (11.3-5b)

Table 19.4.1 shows the parameters of the matched toryces composing the imaginary singulatrons of the ordinary matter level M2 ($m = 1$) in several excitation quantum states n. In the singulatrons, the rotational velocities of the trailing strings of positive and negative toryces have opposite signs, and they are much greater than velocity of light ($\beta_{2r} >> 1$).

Table 19.4.1. Compositions and properties of excited singulatrons of the ordinary matter level $M = 2$ ($m = 1$) in excitation quantum states n.

Singulatrons		Toryces of excited singulatrons			
N	Symbols	Type	b_1	β_{2r}	f_2, Hz
0	$\breve{a}^0_{1,0,0}$	$\breve{A}^-_{1,0,0}$	1.00000000	0.00	3.3864×10^{22}
		$\breve{A}^+_{1,0,0}$	1.00000000	0.00	3.3864×10^{22}
1	$\breve{a}^0_{1,1,0}$	$\breve{A}^-_{1,1,0}$	-0.00366300	274.00	-9.2449×10^{24}
		$\breve{A}^+_{1,1,0}$	0.00363636	-274.00	9.3126×10^{24}
2	$\breve{a}^0_{1,2,0}$	$\breve{A}^-_{1,2,0}$	-0.00182815	548.00	-1.8524×10^{25}
		$\breve{A}^+_{1,2,0}$	0.00182149	-548.00	1.8591×10^{25}
3	$\breve{a}^0_{1,3,0}$	$\breve{A}^-_{1,3,0}$	-0.00121803	822.00	-2.7802×10^{25}
		$\breve{A}^+_{1,3,0}$	0.00121507	- 822.00	2.7870×10^{25}

Table 19.4.2 of shows compositions and properties of excited singulaton helyces emitted by excited singulatrons of the ordinary matter level M2 ($m = 1$). The negative and positive helyces propagate in the opposite directions, forming respectively negative and positive singulatons, as shown in Figure 19.4.

Table 19.4.2. Parameters of the singulaton helyces emitted by excited singulatrons of the ordinary matter level $M = 2$ ($m = 1$).

Singulatons	Helyces of excited singulatons			
Symbols	Symbols	$\tilde{b}_1$	$\tilde{\beta}_{2t}$	$\tilde{f}_1$, Hz
$_0\tilde{a}_{1,0,0}^{1,1,0}$	$-\tilde{A}_{1,0,0}^{1,1,0}$	2.6647×10^{-05}	-3.753×10^{05}	-1.271×10^{27}
	$+\tilde{A}_{1,0,0}^{1,1,0}$	-2.6647×10^{-05}	3.753×10^{05}	1.271×10^{27}
$_0\tilde{a}_{1,1,0}^{1,2,0}$	$-\tilde{A}_{1,1,0}^{1,2,0}$	3.8067×10^{-06}	-2.627×10^{05}	-8.896×10^{27}
	$+\tilde{A}_{1,1,0}^{1,2,0}$	-3.8067×10^{-06}	2.627×10^{05}	8.896×10^{27}
$_0\tilde{a}_{1,2,0}^{1,3,0}$	$-\tilde{A}_{1,2,0}^{1,3,0}$	1.4025×10^{-07}	7.130×10^{05}	-2.415×10^{28}
	$+\tilde{A}_{1,2,0}^{1,3,0}$	-1.4025×10^{-07}	-7.130×10^{05}	2.415×10^{28}

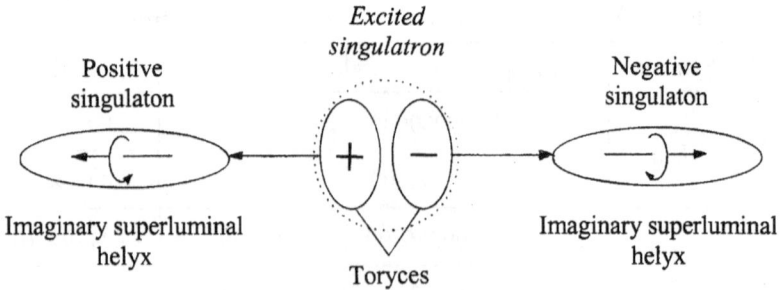

Figure 19.4. Creation of singulatons helyces.

The main features of the emitted singulatons are:

- Translational velocities of singulatons are superluminal, exceeding the velocity of light. When a parental singulatron is transferred from the quantum state $n_k = 2$ to $n_j = 1$, the velocity of emitted singulatron exceeds the velocity of light 2.627×10^{05} times. This velocity increases with an increase in the quantum states n of parental singulatrons.

- Radii of leading strings of singulatons are extremely small. When a parental singulatron is transferred from the quantum state $n_k = 2$ to $n_j = 1$, the relative radius of leading string of the emitted singulatron $b_1 = 3.8067 \times 10^{-06}$. This radius decreases with an increase in the quantum states n of parental singulatrons.

- Frequencies of leading and trailing strings of singulatons are extremely high. When a parental singulatron is transferred from the quantum state $n_k = 2$ to $n_j = 1$, the frequency of emitted singulatron $\tilde{f}_1 = \tilde{f}_1 = 8.896 \times 10^{27}$ Hz. This frequency increases with an increase in the quantum states n of parental singulatrons.

CHAPTER 20

BASIC CONCEPT OF MACRO-TORYX

In the previous chapters we described a role of micro-toryces in the formation of matter particles of the micro-world, including atoms. Since all entities of the macro-world, including tiny grains of sands and gigantic galaxies, are made up of atoms, one may readily conclude that these entities are merely assemblies of micro-toryces.

According to our theory, the entities of the macro-world do not exist alone. Associated with each entity of the macro-world are the spiral spacetimes called *macro-toryces*. Structurally, the macro-toryces are the same as the micro-toryces. The spacetime properties of both of them are based on the same fundamental spacetime equations (the "toryx sacred code") described in Chapter 5. The only difference between them is in the equations defining the radii of their eye radius r_0 that is equal to the radius of real inversion toryx.

20.1 Spacetime Parameters of Macro-Toryx - Figure 20.1 shows the macro-toryx associated with the body A orbited by the body B.

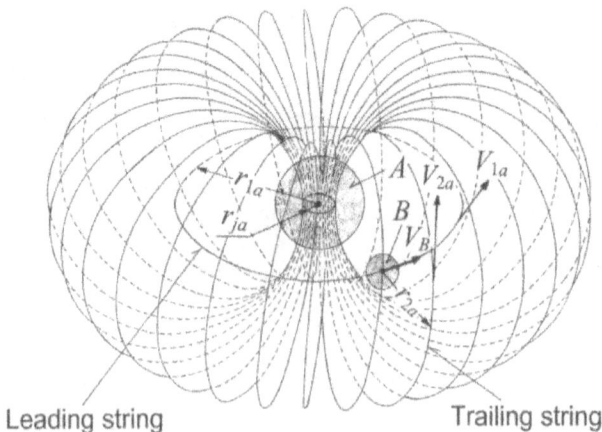

Leading string   Trailing string

Figure 20.1. Central body A with a macro-toryx orbited by the satellite body B.

The macro-toryx contains two strings, a *leading string* and a *trailing string*. The leading string is double-circular; it is moving at the velocity V_{1a} along a circle with the radius r_{1a}. The trailing string is double-toroidal with the radius r_{2a}. Its each toroidal branch accompanies one of the circular branches of the double-circular leading string. Both branches of the toryx trailing string propagate along their toroidal spiral paths at the same spiral velocity V_{2a}.

Spacetime parameters of the macro-toryx are based on the same fundamental spacetime equations (5.3-1) – (5.3-3) applied to the micro-toryx. They are applicable to the range of the radius of leading string radius r_{1a} extending from negative to positive infinity.

- The length of one winding of trailing string L_{2a} is equal to the length of one winding of leading string L_{1a}:

$$L_{2a} = L_{1a} \tag{20.1-1}$$

- The macro-toryx eye radius is equal to the radius of the real inversion macro-toryx r_{ja} and it is constant:

$$r_{ja} = r_{1a} - r_{2a} = const. \tag{20.1-2}$$

- At each point of spiral path of trailing string its spiral velocity V_{2a} is constant and equal to the velocity of light c:

$$V_{2a} = \sqrt{V_{2ta}^2 + V_{2ra}^2} = c = const. \tag{20.1-3}$$

In equation (20.1-3):
V_{2ta} = translational velocity of trailing string
V_{2ra} = rotational velocity of trailing string.

It follows from Eqs (20.1-1) – (20.1-3) that the spiral velocity of leading string V_{1a} is equal to:

$$\beta_{1a} = \frac{V_{1a}}{c} = \frac{\sqrt{2b_{1a} - 1}}{b_{1a}} \tag{20.1-4}$$

Equation (20.1-4) expresses the unified law of planetary motion. This equation is the same as Eq. (10.1-2) except for the relative radius of leading string b_{1a} that is a ratio of the radius of leading string of the macro-toryx r_{1a} to the radius of the real inversion macro-toryx r_{ja}.

$$b_{1a} = \frac{r_{1a}}{r_{ja}} \tag{20.1-5}$$

Notably, the spacetime properties of the macro-toryces are described by the same spacetime equations that are applied to the micro-toryces shown in Table 5.7.

20.2 Derivation of Kepler's Third Law _– One of the most interesting surprises of our theory is that it predicts the Kepler's Third Law of planetary motion. This law states that the cubes of the mean distances of the planets from the Sun r_{1a} are proportional to the squares of their periods of revolution T_{1a}, thus:

$$r_{1a}^3 = kT_{1a}^2 \tag{20.2-1}$$

where k is a constant.

Let us show that the Kepler's Third Law of planetary motion is merely a particular case of the unified law of planetary motion described by Eqs. (20.1-4) and (20.1-5). For the case when $b_{1a} \gg 1$, these equations reduce to the form:

$$V_{1a} = c\sqrt{\frac{2r_{ja}}{r_{1a}}} \tag{20.2-2}$$

Considering that the orbital velocity of the body B around the Sun is equal to:

$$V_{1a} = \frac{2\pi r_{1a}}{T_{1a}} , \tag{20.2-3}$$

we obtain from Eqs. (20.2-2) and (20.2-3) the expression for the Kepler's Third Law in the form:

$$r_1^3 = \frac{r_j c^2}{2\pi^2} T_{1a}^2 \tag{20.2-4}$$

After comparing Eq. (20.2-4) with Eq. (20.2-1), we find that they are identical when the constant k is equal to:

$$k = \frac{r_{ja}c^2}{2\pi^2}$$

(20.2-5)

20.3 Radius of Real Inversion Macro-Toryx – The radius of the real inversion macro-toryx r_{ia} is one of the major parameters allowing us to establish a correlation between physical and spacetime properties of the macro-toryces.

It is possible to express the radius r_{ja} in physical terms by comparing the orbital velocity of the body B circling around the body A based on classical and unified laws of planetary motion.

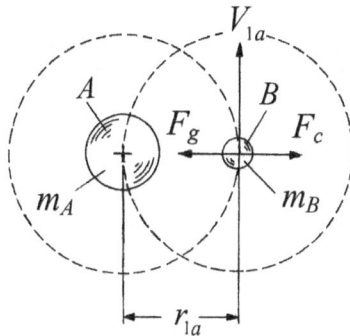

Figure 20.3. Forces applied to a satellite body in a two-body planetary system according to classical mechanics.

The law of planetary motion based on classical mechanics can be derived by equating two forces (Fig. 20.3):

- The attraction gravitational force F_g between the bodies A and B with the masses m_A and m_B separated by the distance r_{1a}

- The centrifugal forces F_c applied to the body B orbiting body A with the orbital velocity V_{1a}.

The gravitational force F_g can be calculated by using Newton's Universal Law of Gravitation. According to this law, the gravitational force F_g between a body with the mass m_A and a body with the mass m_B separated by the distance r_{1a} is equal to:

$$F_g = \frac{m_A m_B G}{r_{1a}^2} \qquad (20.3\text{-}1)$$

The Dutch physicist and astronomer Christiaan Huygens discovered that the centrifugal force F_c applied to the body with the mass m_B orbiting the body m_A with the orbital velocity V_{1a} is equal to:

$$F_c = -\frac{m_B V_{1a}^2}{r_{1a}} \qquad (20.3\text{-}2)$$

A body with the mass m_A will be in equilibrium state with the body m_B when the sum of the gravitational and centrifugal forces is equal to zero. Thus, we obtain from Eqs. (20.3-1) and (20.3-2) that in the equilibrium state, the orbital velocity V_{1a} is inversely proportional to the square root of the orbital radius r_{1a}:

$$V_{1a} = \sqrt{\frac{m_A G}{r_{a1}}} \qquad (20.3\text{-}3)$$

Eq. (20.3-3) expresses the law of planetary motion based on classical mechanics. It is possible to simplify this equation by presenting both the orbital velocity V_{1a} and the orbital radius r_{1a} in relative terms. In respect to the velocity of light c the relative orbital velocity β_{1a} is equal to:

$$\beta_{1a} = \frac{V_{1a}}{c} \qquad (20.3\text{-}4)$$

The relative orbital radius b_{1a} is expressed by Eq. (20.1-5). Consequently, Eq. (20.3-3) transforms into the equation expressing parameters of the classical law of planetary motion in relative terms:

$$\beta_{1a} = \sqrt{\frac{m_A G}{2c^2} \frac{2}{b_{1a} r_{ja}}} \qquad (20.3\text{-}5)$$

For the case when the radius r_{ja} is equal to:

$$r_{ja} = \frac{m_A G}{2c^2} \qquad (20.3\text{-}6)$$

Eq. (20.3-5) reduces to the classical law of planetary motion expressed below in relative values:

$$\beta_{1a} = \sqrt{\frac{2}{b_{1a}}}$$ (20.3-7)

20.4 Unified and Classical Laws of Planetary Motion – After comparing Eq. (20.3-7) based on classical mechanics with Eq. (20.1-4) based on our theory, we find that they are identical for the case when $b_{1a} \gg 1$.

Therefore, the classical law of planetary motion is merely a particular case of the Unified Law of Planetary Motion. Figure 20.4 shows the plots of Eqs. (20.3-7) and Eq. (20.1-4) representing respectively classical and unified laws of motion. The highlights of the plots are:

- Classical law of planetary motion is applied to the range of b_{1a} extending from zero to positive infinity; all values of β_{1a} are expressed with real numbers.
- Unified Law of Planetary Motion is applied to the range of b_{1a} extending from negative to positive infinity; within the range of b_{1a} extending from 0.5 to positive infinity the values of β_{1a} are expressed with real numbers, while within the remaining range of b_{1a} with imaginary numbers.
- When $b_{1a} > 5$, the difference between the calculated values of β_{1a} from the unified and classical laws of planetary motion becomes small and it continues to decrease as b_{1a} increases.
- As b_{1a} decreases from 5 to 2, this difference progressively increases.
- As b_{1a} decreases from 2 to 0, according to the classical law of motion, β_{1a} sharply increases within this range and approaches positive infinity $(+\infty)$.
- As b_{1a} decreases from 2 to 0, according to the unified law of planetary motion, β_{1a} initially increases within the same range and then, after reaching its maximum value of 1 (corresponding to the velocity of light c) at $b_{1a} = 1$, it sharply decreases and approaches negative imaginary infinity $(+0i)$.

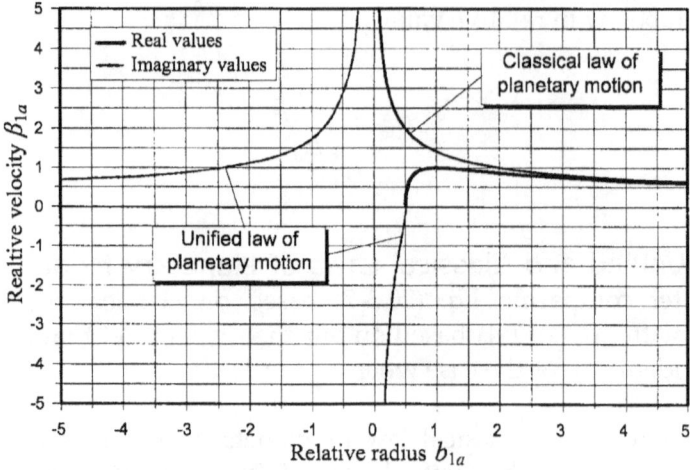

Figure 20.4. Unified law of planetary motion versus classical law of planetary motion as a function of the relative radius b_{1a}.

20.5 Unified Law of Gravitational Force

— According to our theory, the gravitational force is applied to a body when its velocity is not in compliance with the unified law of planetary motion.

Consider a case when the satellite body B moves around the body A along a circular path with the radius r_{1a} at the orbital velocity V_B as shown in Figure 20.1. The body B is embraced by the macro-toryx associated with the central body A in such a way that both the satellite body B and the leading string of the macro-toryx follow the same circular path.

The behavior of the satellite body B depends on a relationship between its orbital velocity V_B and the spiral velocity of leading string V_{1a} of the macro-toryx associated with the central body A. When these velocities are not exactly the same, it means that the motion of the satellite body B does not obey the unified law of planetary motion described by Eq. (20.1-4). The body B will move in a radial direction either towards to or away from the body A with the instant acceleration a_b equal to:

$$a_b = \frac{V_{1a}^2 - V_B^2}{r_{1a}} = \frac{V_{1a}^2(1 - \gamma_V^2)}{r_{1a}} \qquad (20.5\text{-}1)$$

where γ_V is the velocity ratio that is equal to:

$$\gamma_V = \frac{V_B}{V_{1a}} \qquad (20.5\text{-}2)$$

Table 20.5. Mean orbital velocities V_B of the planets of the solar system and the velocities of propagation of the leading string V_{1a} of the macro-toryx associated with the Sun.

Planets	V_{1a}	V_B	V_{1a}/V_B
Mercury	47880.0	47890.0	0.9998
Venus	35026.5	35030.0	0.9999
Earth	29789.6	29783.0	1.0002
Mars	24133.3	24130.0	1.0001
Jupiter	13060.1	13060.0	1.0000
Saturn	9645.3	9640.0	1.0006
Uranus	6801.7	6810.0	0.9988

Table 20.5 shows that the mean orbital velocities V_B of the planets of our solar system are very close to the velocity of propagation of the leading string V_{1a} of the macro-toryx associated with the Sun.

It follows from Eq. (20.5-1) that when $V_B > V_{1a}$ the body B will move away from the central body A with negative acceleration. Conversely, when $V_B < V_{1a}$ the body B will move towards the central body A with positive acceleration. Importantly, the acceleration a_b calculated from Eq. (20.5-1) is applied to the center of gravity of the body C

As long as the satellite body B moves freely either towards to or away from the body A with the acceleration a_b, it will not experience any force applied to it until after its free motion is interrupted by an external obstruction. In the case when the rotational motion of the body B in respect to the body A is completely stopped, there will be a holding force applied to the body B from the external obstruction. We called this force the *unified gravitational force* F_{gu} described by the equation:

$$F_{gu} = m_B a_b \qquad (20.5\text{-}3)$$

Considering that the relative distance b_{1a} between the bodies A and B is equal to:

$$b_{1a} = \frac{r_{1a}}{r_{ja}} \quad , \tag{20.5-4}$$

we obtain from Eqs. (20.1-4), (20.3-6), (20.5-1) – (20.5-4) that the acceleration a_b is equal to:

$$a_b = \frac{m_A G}{r_{1a}^2} \frac{2b_{1a} - 1}{2b_{1a}} (1 - \gamma_V^2) \tag{20.5-5}$$

We find from Eqs. (20.5-3) and (20.5-5) that the unified gravitational force F_g is equal to:

$$F_{gu} = \frac{m_A m_B G}{r_{1a}^2} \frac{2b_{1a} - 1}{2b_{1a}} (1 - \gamma_V^2) \tag{20.5-6}$$

Eq. (20.5-6) describes the *Unified Law of Gravitational Force*. For a particular case when $b_{1a} \gg 1$ and $\gamma_V = 0$, Eq. (20.5-6) reduces to the classical equation for the gravitational force F_g described by Eq. (20.3-1). Thus, Newton's Law of Gravitational Force is a particular case of the unified universal law of gravitational force.

Considering Eq. (20.5-4), we can rewrite Eqs. (20.3-1) and (20.5-6) describing respectively the unified and classical laws of gravitational force in the forms:

$$F_g = \frac{m_A m_B G}{r_{ja}^2} \frac{1}{b_{1a}^2} \tag{20.5-7}$$

$$F_{gu} = \frac{m_A m_B G}{r_{ja}^2} \frac{2b_{1a} - 1}{2b_{1a}^3} (1 - \gamma_V^2) \tag{20.5-8}$$

Based on Eqs. (20.5-7) and (20.5-8), the relative classical and unified gravitational forces f_g and f_{gu} are respectively equal to:

$$f_g = F_g \frac{r_{ja}^2}{m_A m_B G} \frac{1}{b_{1a}^2} \tag{20.5-9}$$

$$f_{gu} = F_{gu} \frac{r_{ja}^2}{m_A m_B G} \frac{2b_{1a} - 1}{2b_{1a}^3} (1 - \gamma_V^2) \tag{20.5-10}$$

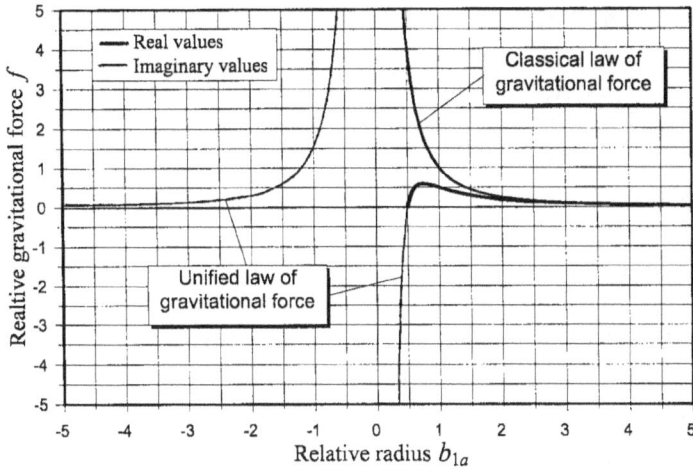

Figure 20.5. Unified Law of Gravitational Force versus Newton's Law of Gravitational Force as a function of the relative radius b_{1a}.

Figure 20.5 shows plots of Eqs. (20.5-9) and (20.5-10) representing Newton's and unified laws of gravitational force for the case when the orbital velocity of the body B around the central body A is equal to zero $(\gamma_V = 0)$. The highlights of the plots are:

- Newton's Law of Gravitational Force is applied to the range of b_{1a} extending from zero to positive infinity; all values of the relative gravitational force f are expressed with real numbers.
- Unified Law of Gravitational Force is applied to the range of b_{1a} extending from negative to positive infinity; within the range of b_{1a} extending from 0.5 to positive infinity the values of f are expressed with real numbers, while within the remaining range of b_{1a} with imaginary numbers.
- When $b_{1a} > 5$, the difference between the calculated values of f from the unified and Newton's laws of gravitational force becomes small and it continues to decrease as b_{1a} increases.
- As b_{1a} decreases from 5 to 2, this difference progressively increases.
- As b_{1a} decreases from 2 to 0, according to Newton's Law of Gravitational Force, f sharply increases within this range and approaches positive infinity $(+\infty)$.

- As b_{1a} decreases from 2 to 0, according to the unified law of gravitational force, f initially increases within the same range and then, after reaching its maximum value, it sharply decreases and approaches negative imaginary in-finility ($+0i$).

Based on Eqs. (20.5-1) – (20.5-10), we may conclude:

- There is no such a phenomenon as "gravitational force" pulling bodies together.
- The "gravitational force" is not applied to the satellite body B when its velocity V_B around the central body A is exactly the same as the velocity of propagation of the leading string V_{1a} of the macro-toryx associated with the central body A.
- The preconditions for appearance of the "gravitational force" are created when the velocity V_B of the satellite body B around the central body A becomes different than the velocity of propagation of the leading string V_{1a} of the macro-toryx associated with the central body. In that case the satellite body B starts to move freely either towards to or away from the body A with the acceleration a_b, but it will not experience any force applied to it.
- The satellite body B will sense the "gravitational force" only after its free motion is interrupted by an external ob-struction.

20.6 Transition From Micro- To Macro-Toryces – According to our theory, the transition from micro- to macro-toryces occurs when the radii of real inversion micro-toryx r_i and macro-toryx r_j are equal to one another, so:

$$r_i = r_j \tag{20.6-1}$$

Let us assume that the gravitational mass of the central body m_A in (20.3-6) is equivalent the gravitational mass of N electrons with the mass m_e each. Consequently, Eq. (20.3-6) will transform into the form:

$$r_{ja} = \frac{N m_e G}{2c^2}, \tag{20.6-2}$$

From Eqs. (10.1-3), (20.6-1) and (20.6-2) we obtain that the transition from micro- to macro-toryces occurs when the gravitational mass of a body will be equal to the mass contained in N_m electrons given by the equation:

$$N_m = \left(\frac{e}{m_e} \right)^2 \frac{1}{4\pi\varepsilon_0 G} ,\qquad (20.6\text{-}3)$$

Considering that: $e = 1.602176565 \times 10^{-19}\ C$,
$m_e = 9.10938291 \times 10^{-31}\ kg$, $\varepsilon_0 = 8.854187817 \times 10^{-12}\ C^2 / N / m^2$ and
$G = 6.674280 \times 10^{-11}\ m^3 / kg / s^2$, we find from Eq. (20.6-3) that
$N_m = 1.020493 \times 10^{21}$.

Depending on the contribution made by the toryces to the interaction between particles the total micro- and macro-worlds can be divided in three groups.

<u>Micro-world</u> – In the micro-world, the interaction between particles is provided by the helyces generated mainly by the micro-toryces. The gravitational mass of each particle m_g is much less than the gravitational mass of N_m electrons with the mass m_e as given by:

$$m_g \ll N_m m_e \qquad (20.6\text{-}4)$$

<u>Intermediate micro-macro-world</u> – In the transitional micro-macro-world, the interaction between particles is provided the helyces generated by both the micro- and macro-toryces. The gravitational mass of each particle m_g is comparable with the gravitational mass of N_m electrons with the mass m_e as given by:

$$m_g \approx N_m m_e \qquad (20.6\text{-}5)$$

<u>Macro-world</u> – In the micro-world, the interaction between particles is provided by the helyces generated mainly by the macro-toryces. The gravitational mass of each particle m_g is much greater than the gravitational mass of N_m electrons with the mass m_e as given by:

$$m_g \gg N_m m_e \qquad (20.6\text{-}6)$$

20.7 Classification of Macro-Toryces – Similarly to the micro-toryces, the macro-toryces are divided into eight groups as a function of the steepness angle of their trailing strings φ_2 as shown in Table 20.7.

As follows from Table 20.7, there are four real macro-toryces corresponding to the range of the steepness angle φ_2 from 0 to 180^0 and four imaginary macro-toryces corresponding to the range of the steepness angle φ_2 from 180 to 360^0. Among four real macro-toryces two are positive and two are negative. Similarly, among four imaginary macro-toryces two are positive and two are negative.

Table 20.7. Classification of macro-toryces.

Groups of macro-toryces	Range of φ_2
Real negative macro-toryces $\langle E \rangle^-_{m,n,q}$	From 0 to 60^0
Real negative macro-toryces $\langle A \rangle^-_{m,n,q}$	From 60 to 90^0
Real positive macro-toryces $\langle A \rangle^+_{m,n,q}$	From 90 to 120^0
Real positive macro-toryces $\langle E \rangle^+_{m,n,q}$	From 120 to 180^0
Imaginary positive macro-toryces $\langle \breve{E} \rangle^+_{m,n,q}$	From 180 to 240^0
Imaginary positive macro-toryces $\langle \breve{A} \rangle^+_{m,n,q}$	From 240 to 270^0
Imaginary negative macro-toryces $\langle \breve{A} \rangle^-_{m,n,q}$	From 270 to 300^0
Imaginary negative macro-toryces $\langle \breve{E} \rangle^-_{m,n,q}$	From 300 to 360^0

In Table 20.7, we show symbols of the macro-toryces inside angular brackets $\langle\ \rangle$ to differentiate them from micro-toryces.

A Path to the Unknown Universe

Similarly to the micro-toryces, the macro-toryces form four kinds of macro-trons:

- Macro-ethertrons
- Macro-singulatrons
- Macro-electrons and
- Macro-positrons.

In their turn, the excited and oscillated macro-trons form respectively excited and oscillated maro-helyces. Consequently, the unified excited and oscillated macro-helyces form the radiation particles of the macro-world respectively called *gravitons* and *gravitrinos*.

The most interesting entities of the macro-world are *black holes*. They are associated with the macro-singulatrons. The black holes are so dense that the radii of their outer bodies are smaller than the radii of their real inversion macro-toryces r_j. Similarly to the singulatons and singulatrinos of the micro-world, the gravitons and gravitrinos propagate with superluminal velocity.

The Universal Spacetime Theory is far from being completed. After twenty years of my analytical research work, writing papers and books and doing lectures on this subject, I still feel that I discovered only a small fraction of secrets hidden inside the toryx. My hope is that other people will also become interested in decoding remaining fascinating secrets of this amazing spacetime entity.

PHYSICAL CONSTANTS

Constant names and symbols		Values
Elementary charge	e	$1.602\,176\,565 \times 10^{-19}$ C
Electric constant	ε_0	$8.854\,187\,817 \times 10^{-12}$ C^2/N/ m^2
Electron mass	m_e	$9.109\,382\,91 \times 10^{-31}$ kg
Proton relative mass	m_p/m_e	$1836.152\,671\,95$
Neutron relative mass	m_n/m_e	$1838.683\,660\,08$
Muon relative mass	m_μ/m_e	$206.768\,284\,26$
Tau relative mass	m_τ/m_e	$3477.151\,011\,54$
Speed of light in vacuum	c	$2.997\,924\,58 \times 10^{08}$ m/s
Newtonian constant of gravitation	G	$6.674\,280 \times 10^{-11}$ m^3/kg/s^2
Planck constant	h	$6.626\,069\,605 \times 10^{-34}$ J s
Inverse fine structure constant	α	$137.035\,999\,11$
Bohr magneton	μ_B	$9.274\,009\,68 \times 10^{-24}$ J/T
Muon magneton	μ_μ	$4.485\,218\,58 \times 10^{-26}$ J/T
Tau magneton	μ_τ	$2.666\,876\,44 \times 10^{-24}$ J/T
Electron magnetic moment to Bohr magneton ratio	μ_e/μ_B	$-1.001\,159\,652\,180\,76$
Muon magnetic moment to muon magneton ratio	μ_μ/μ_μ	$-1.001\,165\,923$
Tau magnetic moment to tau magneton ratio	μ_τ/μ_τ	$-0.987\,629\,486$
Nuclear magneton	μ_N	$5.050\,783\,54 \times 10^{-27}$ J/T
Proton magnetic moment to nuclear magneton ratio	μ_p/μ_N	$2.792\,847\,356$
Neutron magnetic moment to nuclear magneton ratio	μ_n/μ_N	$-1.913\,042\,72$

LIST OF PUBLICATIONS CONSULTED

A

Agassi, J., *Faraday as a Natural Philosopher*, The University of Chicago Press, Chicago, ILL, 1971.

Aiton, E.J., *The Vortex Theory of Planetary Motions*, American Elsevier, Inc., New York, 1972.

Aiton, E.J., *Leibniz - A Biography*, Adam Hilger Ltd, Bristol and Boston, 1985.

Akimov, A.E. and Shipov, G.I., "Torsion Fields and Their Experimental Manifestations," *Proc. of the Int'l Conference on New Ideas in Natural Sciences*, St. Petersburg, Russia, June 1996.

Akimov, A.E. and Tarasenko, V.Y., "Models of Polarized States of the Physical Vacuum and Torsion Fields," *Fizika*, No. 3, March 1992.

Albert, D.Z., "Bohm's Alternative to Quantum Mechanics," *Scientific American*, May 1994.

Alexandersson, O., *Living Water - Victor Schauberger and the Secrets of Natural Energy*, Gateway Books, Bath, UK, 1996.

Alfven, H., *Worlds - Antiworlds: Antimatter in Cosmology*, W.H. Freeman and Company, San Francisco, 1966.

Allen, R.E., *Greek Philosophy: Thales to Aristotle*, The Free Press, New York, 1966.

Andrade, E.N. da C., *Rutherford and the Nature of the Atom*, Doubleday & Company, Inc., New York, 1964.

Arp, H., *Seeing Red - Redshifts, Cosmology and Academic Science*, Apeiron, Montreal, Canada, 1998.

Ash, D. and Hewitt, P., *The Vortex - Key to Future Science*, Gateway Books, Bath, England, 1991.

B

Babbitt, E.D., *The Principles of Light and Color*, Babbitt & Co., New York, 1878.

Baggott, J., *The Meaning of Quantum Theory*, Oxford University Press, Oxford, UK, 1992.

Barrow, J.D. and Silk, J., *The Left Hand of Creation*, Oxford University Press, New York, 1983.

Barrow, J.D., *The Constants of Nature – The Numbers that Encode the Deepest Secrets of the Universe*, Vintage Books A Division of Random House, Inc., New York, 2002.

Barrow, J.D., *The Infinite Book – A Short Guide to the Boundless, Timeless and Endless*, Pantheon Books, New York, 2005.

Barrow, J.D., *New Theories of Everything – The Quest for Ultimate Explanation*, Oxford University Press Inc., New York, 2007.

Bartusiak, M., "Loops of Space," *Discover*, April 1993.

Bartusiak, M., "Gravity Wave Sky," *Discover*, July 1993.

Beckmann, P., *A History of PI*, Dorset Press, New York, 1989.

Bhadkamkar, A. and Fox, H., "Electron Charge Cluster Sparking in Aqueous Solutions," *Journal of New Energy*, Vol. 1, No. 4, 1996.

Beiser, G., *The Story of Gravity - An Historical Approach to the Study of the Force That Holds the Universe Together*, E.P. Dutton & Co., Inc., New York, 1968.

Bekenstein, J.D., "Information in the Holographic Universe," *Scientific American*, August 2003, pp. 59-65.

Bergman, D.L. and Wesley, J.P., "Spinning Charge Ring Model of Electron Yielding Anomalous Magnetic Moment," *Galilean Electrodynamics*, Vol. 1, No. 5, Sept./Oct. 1990.

Blackwood, O.H., et al, *An Outline of Atomic Physics*, John Wiley & Sons, Inc., New York, 1955.

Bloyd, J.G., *Broken Arrow of Time - Rethinking the Revolution in Modern Physics*, Writers Club Press, San Jose, CA, 2001.

Born, M., *Atomic Physics*, Dover Publications, Inc., Mineola, NY, 1989.

Boscovich, R.J., *A Theory of Natural Philosophy*, The M.I.T. Press, Cambridge, MA, 1966.

Boslough, J., *Stephen Hawking's Universe - An Introduction to the Most Remarkable Scientist of Our Time*, Quill/William Morrow, New York, 1985.

Boslough, J., *Masters of Time - Cosmology at the End of Innocence*, Addison-Wesley Publishing Company, Reading, MA, 1992.

Bostick, W., "Mass, Charge, and Current: The Essence of Morphology," *Physics Essays*, Vol. 4, No. 1, pp. 45-59, March 1991.

Bowers, B., *Michael Faraday and Electricity*, Priory Press Ltd., London, 1974.

Broglie, L., de, *The Revolution in Physics*, The Noonday Press, New York, 1953.

Broglie, L., de, *New Perspectives in Physics*, Basic Books, Inc. Publishers, New York, 1962.

Burger, T.J., *Nature* **271**, 402, 1978.

C

Calladine, C.R. and Drew, H.R., *Understanding DNA – The Molecule and How It Works*, Second Edition, Academic Press, New York, 2002.

Cambier, J-L. and Micheletti, D.A., "Theoretical Analysis of the Electron Spiral Toroidal Concept," NASA/CR-2000-210654, Dec. 2000.

Capra, F., *The Web of Life*, Anchor Books/Random House, Inc., 1997.

Capra, F., *The Tao of Physics*, Shambhala Publications, Inc., 1999.

Carter, J., *The Other Theory of Physics - A Non-Field Unified Theory of Matter and Motion*, Absolute Motion Press, 2000.

Cartledge, P., *Democritus*, Routledge, New York, 1999.

Caspar, M., *Kepler*, Abelard-Schuman, London and New York, 1992.

Cecil, T.E. and Chern, S., *Tight and Taut Submanifolds*, Cambridge University Press, New York, 1997.

Chalidze, V., *Mass and Electric Charge in the Vortex Theory of Matter*, Universal Publishers, 2001.

Clark, G., *The Man Who Tapped the Secrets of the Universe*, The University of Science and Philosophy, Swannanoa, Waynesboro, 2000.

Clawson, C.C., *Mathematical Sorcery - Revealing the Secrets of Numbers*, Perseus Publishing, Cambridge, MA, 1999.

Close, F., Marten, M., & Sutton, C., *The Particle Explosion*, Oxford University Press, New York, 1994.

Coats, C., *Living Energies*, Gateway Books, Bath, UK, 1996.

CODATA Recommended Values, *The NIST Reference on Constants, Units and Uncertainties*, 2011.

Cook, N., *The Hunt for Zero Point - Inside the Classified World of Antigravity Technology*, Broadway Books, New York, 2001.

Cook, T.A., *The Curves of Life*, Dover Publications, Inc., New York, 1979.

Coxeter, H.S.M., *Introduction to Geometry*, John Wiley & Sons, Inc., New York, 1961.

Coxeter, H.S.M., *The Beauty of Geometry*, Dover Publications, Inc., New York, 1968.

Crew, H., *The Wave Theory of Light - Memoirs by Huygens, Young and Fresnel*, American Book Company, New York, 1900.

D

Dalton, J., et al, *Foundations of the Atomic Theory: Comprising Papers and Extracts*, Alembic Club, Edinburgh, UK, 1968.

Davies, P.C.W. and Brown, J., *Superstrings - A Theory of Everything?*, Cambridge University Press, Cambridge, UK, 1988.

Davies, P. and Gribbin, J., *The Matter Myth - Dramatic Discoveries That Challenge Our Understanding of Physical Reality*, Touchstone Book/Simon & Schuster, New York, 1992.

Davies, P., *About Time - Einstein's Unfinished Revolution*, Simon & Schuster, New York, 1995.

Day, W., *Bridge from Nowhere - The Photonic Origin of Matter*, Rhombics, Cambridge, MA, 1996.

Day, W., *A New Physics - Foundation for New Directions*, Cambridge, MA, 2000.

Derbyshire, J., *Unkn( )wn Quantity - A Real and Imaginary History of Algebra*, A Plume Book, Published by Penguin Group, New York, 2007.

Di Mario, D., "Electrogravity: A Basic Link Between Electricity and Gravity," *Speculations in Science and Technology*, Vol. 20, No. 4, Dec. 1997.

Dibner, B., *Oersted - And the Discovery of Electromagnetism*, Blaisdell Publishing Company, New York, 1962.

Dijksterhuis, E.J., *Archimedes*, Princeton University Press, Princeton, N.J., 1987.

Dmitriyev, V.P., "Mechanical Analogy for the Wave - Particle: Helix on Vortex Filament," *Apeiron*, Vol. 8, No. 2, April 2001.

Domb, C., *Clerk Maxwell and Modern Science - Six Commemorative Lectures*, The Athlone Press, University of London, UK, 1963.

Drake, S., *Galileo at Work, His Scientific Biography*, The University of Chicago Press, Chicago, 1978.

Drew, H.R., "The Electron as a Four-Dimensional Helix of Spin-1/2 Symmetry,"

Physics Essays, Vol. 12, No. 4, 1999.

Driscoll, R.B., *United Theory of Ether, Field and Matter*, Published by Author, Portland, OR, 1964.

Driscoll, R.B., *United Theory of Ether, Field and Matter (supplement)*, Published by Author, Oakland, CA, 1965.

Duncan, J.C., *Astronomy – A Textbook*, Fifth Edition, Harper & Brothers Publishers, New York, 1926.

E

Eckhart, L., *Four-Dimensional Space*, Indiana University Press, Bloomington, 1968.

Edwards, E.B., *Pattern and Design with Dynamic Symmetry*, Dover Publications, Inc., New York, 1932.

Edwards, L., *The Vortex of Life – Nature's Patterns in Space and Time*, Floris Books, Edinburgh, UK, 2006.

Einstein, A. and Hopf, L., Ann. Phys., 33, 1096 (1910a):Ann. Phys., 33, 1105 (1910b).

Einstein, A., *Out of My Later Years*, A Citadel Press Book-Carol Publishing Group, New York, NY, 1991.

Einstein, A., *Relativity - The Special and the General Theory*, Crown Publishers, Inc., New York, 1961.

Einstein, A., Infeld, L., *The Evolution of Physics - From Early Concepts to Relativity and Quanta*, A Touchstone Book/Simon & Schuster, New York, 1966.

Einstein, A., "Aether and the Theory of Relativity," (Address on May 5, 1920, at the University of Leyden), *Journal of New Energy*, Vol. 7, No. 1, 2003.

Elgin, D., *The Living Universe – Where are We? Who are We? Where are We Going?*, Berrett-Koehler Publishers, Inc., San Francisco, CA, 2009.

Epstein, L.C., *Thinking Physics Is Gedanken Physics*, Insight Press, San Francisco, CA, 1983.

Epstein, L.C., *Relativity Visualized*, Insight Press, San Francisco, CA, 1992.

F

Farrington, B., *Greek Science - Its Meaning for Us*, Penguin Books, Baltimore, MD, 1971.

Ferguson, K., *Stephen Hawking - Quest for a Theory of Everything*, Bantam Books, New York, 1992.

Flood, R. and Lockwood, M., *The Nature of Time*, Basil Blackwell, Inc., Cambridge, MA, 1990.

Folger, T., "Tangled Up In Strings – Two Books Say That Today's Theoretical Physicists Are Way Off Course," *Discover*, Sept. 2006.

Ford, K.W., *101 Quantum Questions*, Harvard University Press, Cambridge, Massachusetts, 2011.

Frank, P., *Einstein - His Life and Times*, Da Capo Press, Inc., New York, 1947.

Fraser, et al, *The Search for Infinity*, Facts on File, Inc., New York, 1995.

Freedman, D.H., "The Mysterious Middle of the Milky Way," *Discover*, November 1998.

Freeman, K. and McNamara, G., *In Search of Dark Matter*, Springer Praxis Publishing, Chichester, UK, 2006.

Friedman, N., *Bridging Science and Spirit - Common Elements in David Bohm's Rhysics, The Perennial Philosophy and Seth*, Living Lake Books, St. Louis, MO, 1994.

Fritzsch, H., *Quarks - The Stuff of Matter*, Basic Books, Inc., New York, 1983.

Fritzsch, H., *The Creation of Matter - The Universe From Beginning to End*, Basic Books, Inc., New York, 1984.

Funk & Wagnalls New Encyclopedia, Funk & Wagnalls, Inc., USA, 1966.

G

Gamow, G., *Thirty Years That Shook Physics - The Story of Quantum Theory*, Dover Publications, Inc., New York, 1985.

Gamow, G., *One, Two, Three ... Infinity - Facts and Speculations of Science*, Dover Publications, Inc., New York, 1988.

Gardner, M., *Knotted Doughnuts and Other Mathematical Entertainments*, W.H. Freeman and Company, New York, 1986.

Gauthier, R., "Faster-than-light quantum models of the photon and the electron", in M. S. El-Genk, (ed.) *"Space Technology and Applications International Forum – STAIF 2007"*, American Institute of Physics 978-0-7354-0386-4/07, p1099-1108, (2007).

Gauthier, R., "Transluminal energy quantum models of the photon and the electron", in R.L. Amoroso, P. Rowlands & L.H. Kauffman (eds.) *The Physics of Reality: Space, Time, Matter, Cosmos, 8th Symposium in Honor of Mathematical Physicist Jean-Pierre Vigier*, Hackensack: World Scientific, (2013).

Gauthier, R., "A transluminal energy quantum model of the cosmic quantum", in R.L. Amoroso, P. Rowlands & L.H. Kauffman (eds.) *The Physics of Reality: Space, Time, Matter, Cosmos, 8th Symposium in Honor of Mathematical Physicist Jean-Pierre Vigier*, Hackensack: World Scientific, (2013).

Gazale, M.J., *Gnomon*, Princeton University Press, Princeton, N.J., 1999.

Gell-Mann, M., *Quark and the Jaguar – Adventures in the Simple and the Complex*, A W, H. Freeman/Owl Book, Henry Holt and Company, LLC, New York, 1994.

Gell-Mann, M., *Complexity*, Vol. 1, no 5, John Wiley and Sons, Inc., New York, 1995/96.

Genz, H., *Nothingness - The Science of Empty Space*, Perseus Books, Reading, MA, 1999.

Geymonat, L., *Galileo Galilei: A Biography and Inquiry Into His Philosophy of Science*, McGraw-Hill Book Company, 1965.

Ghosh, A., *Origin of Inertia*, Apeiron, Montreal, Canada, 2000.

Ghyka, M., *The Geometry of Art and Life*, Dover Publications, Inc, New York, 1977.

Gillispie, C.C., *Dictionary of Scientific Biography*, Vol. IV, Charles Scribner's Sons, New York, 1971.

Ginzburg, V.B., "Toroidal Spiral Field Theory," *Speculations in Science and*

Technology, Vol. 19, 1996.

Ginzburg, V.B., *Spiral Grain of the Universe - In Search of the Archimedes File*, University Editions, Inc., Huntington, WV, 1996.

Ginzburg, V.B., "Structure of Atoms and Fields," *Speculations in Science and Technology*, Vol. 20, 1997.

Ginzburg, V.B., "Double Helical and Double Toroidal Spiral Fields," *Speculations in Science and Technology*, Vol. 22, 1998.

Ginzburg, V.B., *Unified Spiral Field and Matter - A Story of a Great Discovery*, Helicola Press, Pittsburgh, PA, 1999.

Ginzburg, V.B., "Nuclear Implosion," *Journal of New Energy*, Vol. 3, No. 4, 1999.

Ginzburg, V.B., *Continuous Spiral Motion System for Rolling Mills*, U.S. Patent No. 5,970,771, Oct. 26, 1999.

Ginzburg, V.B., *Continuous Spiral Motion System and Roll Bending System for Rolling Mills*, U.S. Patent No. 6,029,491, Feb. 29, 2000.

Ginzburg, V.B., "Dynamic Aether," *Journal of New Energy*, Vol. 6, No. 1, 2001.

Ginzburg, V.B., *The Unification of Strong, Gravitational & Electric Forces*, Helicola Press, Pittsburgh, PA, 2003.

Ginzburg, V.B., "Electric Nature of Strong Interactions," *Journal of New Energy*, Vol. 7, No. 1, 2003.

Ginzburg, V.B., "Unified Spiral Field Theory – A Quiet Revolution in Physics," *VIA-Vision in Action*, Vol. 2, No. 1 & 2, 2004.

Ginzburg, V.B., "The Relativistic Torus and Helix as the Prime Elements of Nature," *Proceedings of the Natural Philosophy Alliance*, Vol. 1, No. 1, Spring 2004.

Ginzburg, V.B., *Prime Elements of Ordinary Matter, Dark Matter & Dark Energy*, Helicola Press, Pittsburgh, PA, 2006.

Ginzburg, V.B., *Prime Elements of Ordinary Matter, Dark Matter & Dark Energy – Beyond Standard Model & String Theory*, The second revised edition, Universal Publishers, Boca Raton, Florida, 2007.

Ginzburg, V.B., "The Unification of Forces," *Proceedings of the Natural Philosophy Alliance*, Vol. 4, No. 1, 2007.

Ginzburg, V.B., "The Origin of the Universe, Part 1: Toryces," *Proceedings of the Natural Philosophy Alliance*, The 17th Annual Conference of the NPA 23-26 June 2010 at California State University Long Beach, Vol. 7.

Ginzburg, V.B., "Basic Concept of 3-Dimensional Spiral String Theory (3D-SST)," *Proceedings of the Natural Philosophy Alliance*, The 18th Annual Conference of the NPA, 6-9 July 2011 at the University of Maryland, College Park, USA, Vol. 8.

Gleick, J., *Chaos - Making a New Science*, Penguin Books, New York, 1988.

Gleick, J., *Genius - The Life and Science of Richard Feynman*, Pantheon Books, New York, 1992.

Gorini, C.A., *Geometry*, Facts On File, Inc., New York, 2003.

Goswami, A., *The Self-Aware Universe - How Consciousness Creates the Material World*, Penquin Putnam Inc., 1993.

Gray, A., *Lord Kelvin - An Account of His Scientific Life and Work*, E.P. Dutton & Co., 1908.

Gray, A., *Modern Differential Geometry of Curves and Surfaces*, CRC Press, Boca Raton, 1993.

Greene, B., *The Elegant Universe - Superstrings, Hidden Dimensions, and the Quest for the Ultimate Theory*, W.W. Norton & Company, New York, 1999.

Greene, B., *The Fabric of the Cosmos*, Alfred A. Knopf, New York, 2004.

Gribbin, J., *Q Is For Quantum, An Encyclopedia of Particle Physics*, A Touchstone Book/ Simon & Schuster, New York, 2000.

H

Haisch, B., *The Purpose-Guided Universe – Believing in Einstein, Darwin, and God*, The Career Press, Inc., Franklin Lakakes, NJ, 2010.

Haisch, B., *The God Theory - Universes, Zero-Point Fields, and What's Behind It All*, Weser Books, San Francisco, CA, 2006..

Hall, A.R., *Isaac Newton - Adventurer in Thought*, Blackwell Publishers, Oxford, UK, 1994.

Hambidge, J., *Practical Applications of Dynamic Symmetry*, The Devin-Adair Company, New York, 1967.

Hargittai, I., Pickover, C.A., *Spiral Symmetry*, World Scientific Publishing Co., Singapore, 1992.

Harrison, L.P., *Meteorology*, National Aeronautics Council, Inc., New York, 1942.

Hawking, S., *Black Holes and Baby Universes and Other Essays*, Bantam Books, New York, 1993.

Hawking, S., *A Brief History of Time - From the Big Bang to Black Holes*, Bantam Books, New York, 1990.

Hawking, S. and Penrose, R., *The Nature of Space and Time*, Princeton University Press, Princeton, NJ, 2000.

Heath, J.L., *The Works of Archimedes*, Dover Publications, Inc., New York, 1953.

Heilbron, J.L., *The Dilemmas of an Upright Man - Max Planck as Spokesman for German Science*, University of California Press, Berkeley, 1986.

Heisenberg, E., *Inner Exile - Recollections of a Life with Werner Heisenberg*, Birkhauser Boston, MA, 1980.

Henderson, L.D., *The Fourth Dimension and Non-Euclidean Geometry in Modern Art*, Princeton, 1983.

Herzberg, G., *Atomic Spectra and Atomic Structure*, Dover Publications, Inc., New York, 1944.

Hey, N., *Solar System*, Weidenfeld & Nicolson, The Orion Publishing Group, Wellington House, London, UK.

Hey, T. and Walters, P., *The New Quantum Universe*, Cambridge University Press, UK, 2003.

Hippel, F., *Citizen Scientist*, A Touchstone Book/Simon & Schuster, New York, 1991.

Hoffmann, B., *Albert Einstein - Creator and Rebel*, New American Library, New York, 1972.

Hoffman, R.N., "Controlling Hurricanes – Can Hurricanes and Other Severe Tropical Storms be Moderated or Deflected?" *Scientific American*, Oct.

2004.

Hooft, G., *In Search of the Ultimate Building Blocks*, Cambridge University Press, UK, 1997.

Horgan, J., "Gravity Quantized? - A Radical Theory of Gravity Weaves Space From Tiny Loops," *Scientific American*, September 1992.

Hotson, D.L., "Dirac's Equation and the Sea of Negative Energy, Part 1," *Infinite Energy*, Vol. 8, Issue 43, 2002.

Hotson, D.L., "Dirac's Equation and the Sea of Negative Energy, Part 2," *Infinite Energy*, Vol. 8, Issue 44, 2002.

I

Icke, V., "From Expansion to Intelligence in the Universe," *Speculations in Science and Technology*, Vol. 14, No. 4, 1991.

Ipsen, D.C., *Archimedes: Greatest Scientist of the Ancient World*, Enslow Publishers, Inc., Hillside, N.J., 1988.

J

Jammer, M., *Concepts of Mass in Classical and Modern Physics*, Dover Publications, Inc., Mineola, New York, 1997.

Jammer, M., *Concepts of Force*, Dover Publications, Inc., Mineola, New York, 1999.

Johnson, G., "The Inelegant Universe – Two New Books Argue That It Is Time For String Theory To Give Way," *Scientific American*, September, 2006.

Jonsson, I., *Emanuel Swedenborg*, Twayne Publishers Inc., New York, 1971.

K

Kafatos, M. and Nadeau, R., *The Conscious Universe - Part and Whole in Modern Physical Theory*, Springer-Verlag New York, Inc., New York, 1990.

Kaku, M., *Introduction to Superstrings*, Springer-Verlag, New York, 1988.

Kaku, M., *Hyperspace*, Anchor Books/Doubleday, New York, 1994.

Kaku, M., *Beyond Einstein - The Cosmic Quest for Theory of the Universe*, Anchor Books/Doubleday, New York, 1995.

Kanarev, F.M., *The Foundation of Physchemistry of Microworld*, Kuban State Agrarian University (KSAU), Krasnodar, Russia, 2002.

Kane, G., *The Particle Garden - Our Universe as Understood by Particle Physicists*, Addison-Wesley Publishing Company, Reading, MA, 1995.

Kaplan, R., *The Nothing That Is - A Natural History of Zero*, Oxford University Press, Oxford, New York, 1999.

Kaplan, R. and Kaplan, E., *The Art of The Infinite*, Oxford University Press, Oxford, New York, 2003.

Kaufmann, W.J., *Black Holes and Warped Spacetime*, W.H. Freeman and Company, San Francisco, CA.

Kimura, Y.G., *Think Kosmically Act Globally*, Contact Printing, North Vancouver, B.C., Canada, 2000.

Kimura, Y.G., *The Book of Balance*, (Translation), The University of Science and Philosophy, Contact Printing, North Vancouver, B.C., Canada, 2002.

Kimura, Y.G., "The Transcendent Unity of Science and Spirituality," *VIA* –

Vision in Action, Vol. 2, No. 1 & 2, 2004.

King, M.B., "Vortex Filaments, Torsional Fields and the Zero-Point Energy," *Journal of New Energy*, Vol. 3, No. 2/3, 1998.

King, M.B., "Dual Vortex Forms: The Key to a Large Zero-Point Energy Coherence," *Journal of New Energy*, Vol. 5, No. 2, 2000.

King, M.B., *Quest for Zero Point Energy*, Adventures Unlimited Press, Kempton, IL, 2001.

Knight, D.C., *The Science Book of Meteorology*, Franklin Watts, Inc., New York, 1964.

Krauss, L.M., *Quintessence - The Mystery of Missing Mass in the Universe*, Basic Books, New York, NY, 2000.

L

Lamb, G.L., "Solutions and the Motion of Helical Curves," *Physical Review Letters*, Vol. 37, No. 5, August 1976.

Lakhtakia, A. and Weiglhofer, W.S., "Time-Dependent Beltrami Fields in Free Space: Dyadic Green Functions and Radiation Potentials," *Physical Review E*, Vol. 49, Number 6, June 1994.

Lakhtakia, A. and Weiglhofer, W.S., "Covariances and Invariances of the Beltrami-Maxwell Postulates," *IEE Proc. - Sci. Meas. Technol.*, Vol. 142, No. 3, May 1995.

Lang, T.G., "Proposed Unified Field Theory – Part II: Protons, Neutrons and Fields," *Galilean Electrodynamics*, Vol. 12, No. 6, Nov./Dec. 2001.

Larsen, R., et al, *Emanuel Swedenborg - A Continuing Vision*, Swedenborg Foundation, Inc., New York, 1988.

Laugwitz, D., *Differential and Riemannian Geometry*, Academic Press, New York, 1965.

Lauwerier, H., *Fractals - Endlessly Repeated Geometrical Figures*, Princeton University Press, Princeton, New Jersey, 1991.

Lederman, L.M. and Teresi, D., *The God Particle*, Bantam Doubleday Dell Publishing Group, Inc., New York, 1993.

Lederman, L. and Hill, C.T., *Symmetry and the Beautiful Universe*, Prometheus Books, New York, 2004.

Lederman, L.M. and Hill, C.T., *Quantum Physics for Poets*, Prometheus Books, New York, 2011.

Lerner, E.J., *The Big Bang Never Happened*, Vintage Books, Random House, Inc., New York, 1992.

Lindgren, C.E., *Four-Dimensional Descriptive Geometry*, McGraw-Hill Book Company, New York, 1968.

Lindley, D., *The End of Physics - The Myth of a United Theory*, HarperCollins Publishers, Inc., 1993.

Livio, M., *The Golden Ratio*, Broadway Books, New York, 2002.

Lockwood, E.H., *A Book of Curves*, Cambridge University Press, New York, 1961.

Lorentz, H.A., *Problems of Modern Physics - A Course of Lectures Delivered in the CA Institute of Technology*, Ginn and Company, Boston, 1927.

Lucas, C.W., "A Classical Electromagnetic Theory of Elementary Particles," *Journal of New Energy*, Vol. 6, No. 4, 2002.

Lucas, C.W. "A Classical Electromagnetic Theory of Elementary Particles Part 2, Interwining Charge-Fibers," The Journal of Common Sense Science, Foundation of Science, May 2005, Vol. 8 No. 2.

Ludwig, C., *Michael Faraday - Father of Electronics*, Herald Press, Scottdale, PA, 1978.

Lugt, H.J., *Vortex Flow in Nature and Technology*, Krieger Publishing Company, Malabar, Florida, 1995.

M

Maldacena, J., "The Illusion of Gravity," *Scientific American*, November 2005, pp. 57-63.

Magueijo, J., *Faster Than the Speed of Light - The Story of A Scientific Speculation*, Penguin Books, New York, 2003.

Manning, J., *The Coming Energy Revolution - The Search for Free Energy*, Avery Publishing Group, Garden City Park, New York, 1996.

Manning, H.P., *The Fourth Dimension Simply Explained*, Dover Publications, Inc., New York, 1960.

Maor, E., *e: The Story of a Number*, Princeton University Press, Princeton, NJ, 1994.

Maor, E., *Trigonometric Delights*, Princeton University Press, Princeton, NJ, 1998.

Marsden, J.E. and McCracken, M., *The HOPF Bifurcation and Its Application*, Springler-Verlag, New York, 1976 (Russian).

Mazur, B., *Imagining Numbers (particularly the square root of minus fifteen)*, Picador, New York, 2003.

McCrea, W.H., "Arthur Stanley Eddington," *Scientific American*, June 1991.

McCutcheon, M., *The Final Theory – Rethinking Our Scientific Legacy*, Universal Publishers, Boca Raton, Florida, 2004.

McLeish, J., *The Story of Numbers*, Fawcett Columbine, New York, 1991.

Millar, D., et al, The Cambridge Dictionary of Scientists, Cambridge University Press, 1996.

Miller, A.I., *137 – Jung, Pauli and the Pursuit of the Scientific Obsession*, W.W Norton & Company, Inc., New York, 2009.

Milton, R., *Alternative Science - Challenging the Myths of the Scientific Establishment*, Park Street Press, Rochester, Vermont, 1996.

Mitsopoulos, T.D., "Similarity Between Elementary Particles and Electric Circuits," *Galilean Electrodynamics*, Vol. 12, No. 6, Nov./Dec. 2001.

Moore, W., *Schrodinger - Life and Thought*, Cambridge University Press, UK, 1992.

Mugnai, D., et al, "Observation of Superluminal Behaviors in Wave Propagation," *Physical Review Letters*, Vol. 84, Number 21, May 2000.

Murchie, G., *The Seven Mysteries of Life - An Exploration in Science and Philosophy*, Houghton Mifflin Company, Boston, 1978.

N

Nahin, P.J., *An Imaginary Tale – The Story of* $\sqrt{-1}$, Princeton University Press, Princeton, New Jersey, 1998.

Nernst, W., Verh. Dtsch. Phys. Ges., 18, 83, 1916.

Newton, I., *The Principia*, Prometheus Books, Amherst, New York, 1995.

Niven, W.D., *The Scientific Papers of James Clerk Maxwell*, Dover Publications, Inc., New York, 1890.

Novikov, I.D., *The River of Time*, Cambridge University Press, Cambridge, UK, 1998.

O

Okun, L.B., "The Concept of Mass," *Physics Today*, Vol. 42, June 1989.

Oliwensrein, L., "Bent out of Shape," *Discover*, July 1993.

Oros di Bartini, R., "Relations Between Physical Constants," *Progress in Physics*, v. 3, October 2005, pp. 34-40.

Oros di Bartini, R., "Some Relations Between Physical Constants," *Doklady* Acad. Nauk USSR, 1965, v. 163, No. 4, pp. 861-864.

Oschman, J.L., *Energy Medicine – The Scientific Basis,* Churchill Livingstone, New York, 2000.

P

Parry, A., *The Russian Scientist*, The Macmillan Company, New York, 1973.

Parson, A.L., "A Magneton Theory of the Structure of the Atom," Smithsonian Miscellaneous Collections, Vol. 65, No. 11, Publication 2371, Nov. 29, 1915.

Pauli, W., *Theory of Relativity*, Pergamon Press, London, UK, 1958.

Peat, F.D., *Superstrings and the Search for The Theory of Everything*, Contemporary Books, Lincolnwood (Chicago), ILL,1988.

Pedoe, D., *Geometry - A Comprehensive Course*, Dover Publications, Inc., New York, 1988.

Penrose, R., *The Road to Reality - A Complete Guide to the Laws of the Universe*, Alfred A. Knopf, New York, 2005.

Peratt, A.L., "Birkeland and the Electromagnetic Cosmology," *Sky & Telescope*, May 1985.

Physical Review D: Particles and Fields, Vol. 54, The American Physical Society, 1996.

Pickover, C.A., *Mathematics and Beauty II; Spirals and "Strange" Spirals in Civilization, Nature, Science, and Art*, IBM Thomas J. Watson Research Center, Yorktown Heights, NY, 1987.

Porter, R., *The Biographical Dictionary of Scientists*, Oxford University Press, New York, 1994.

Price, W.C., et al, *Wave Mechanics; The First Fifty Years - A Tribute to Professor Louis De Broglie*, John Wiley & Sons, New York-Toronto, 1973.

Price, H., *Time's Arrow and Archimedes' Point*, Oxford University Press, New York, 1996.

Purdy, S., and Sandak, C.R., *Ancient Greece*, Franklin Watts, New York, 1982.

Puthoff, H.E., et al, "Engineering the Zero-Point Field and Polarizable Vacuum for Interstellar Flight," *Journal of New Energy*, Vol. 6, No. 1, 2001.

R

Reed, D., "Excitation and Extraction of Vacuum Energy Via EM-Torsional Field Coupling - Theoretical Model," *Journal of New Energy*, Vol. 3, No. 2/3, 1998.

Reed, D., "A New Paradigm for Time – Evidence From Empirical and Esoteric Sources," *Journal of New Energy*, Vol. 6, No. 2, 2001.

Ridley, B.K., *Time, Space and Things*, Cambridge University Press, Cambridge, UK, 1994.

Riordan, M., *The Hunting of the Quark - A True Story of Modern Physics*, Simon and Schuster/Touchstone, New York, 1987.

Riordan, M. and Schramm, D.N., *The Shadows of Creation - Dark Matter and the Structure of the Universe*, W.H. Freeman and Company, New York, 1991.

Rucker, R., *The Fourth Dimension*, Houghton Mifflin Company, Boston, 1984.

Russell, P., *The White Hole in Time*, Harper San Francisco, 1992.

W. Russell, *The Universal One*, University of Science and Philosophy, Swannanoa, Waynesboro, Virginia, 1974.

Russell, W., *A New Concept of the Universe*, The University of Science and Philosophy, Swannanoa, Waynesboro, VA, 1989.

Russell, W., *The Secret of Light*, The University of Science and Philosophy, Swannanoa, Waynesboro, VA, 1994.

Ryu, C., *The Grand Unified Theory – A Scientific Theory of Everything*, PublishAmerica, Baltimore, 2004.

S

Salem, K.G., *The New Gravity - A New Force - A New Mass - A New Acceleration - Unifying Gravity with Light*, Salem Books, Johnstown, PA, 1994.

Sanders, P.A. Jr., *Scientific Vortex Information*, Free Soul Publishing, Sedona, AZ, 1992.

Sano, C., "Twisting & Untwisting of Spirals of Aether and Fractal Vortices Connecting Dynamic Aethers," *Journal of New Energy*, Vol. 6, No. 2, 2001.

Sarg, S., "A Physical Model of the Electron According to the Basic Structure of Matter Hypothesis," *Physics Essays*, Vol. 16, No.2, 180-195, 2003.

Sarg, S, "Basic Structure of Matter – Supergravitation Unified Theory Based on an Alternative Concept of the Physical Vacuum," Proceedings of the 17th Annual Conference of the NPA, 23-26 June, 2010, at Long Beach, CA, Vol. 7, 2010, pp. 479-484.

Savov, E., *Theory of Interaction - The Simplest Explanation of Everything*, Geones Books, Sofia, Bulgaria, 2002.

Schneider, M.S., *A Beginner's Guide to Constructing the Universe*, HarperPerennial, New York, 1995.

Schwerdtfeger, H., *Geometry of Complex Numbers – Circle Geometry, Moebius Transformation, Non-Euclidean Geometry*, Dover Publication, Inc., New York, 1979.

Scientific American, *The Enigma of Weather*, Scientific American, New York, NY, Oct. 2004.

Seggern, D.H. von, *CRC Handbook of Mathematical Curves and Surfaces*, CRC Press, Boca Raton, Florida, 1990.

Seggern, D.H. von, *CRC Standard Curves and Surfaces*, CRC Press, Boca Raton, Florida, 1993.

Segre, E., *Nuclei and Particles - An Introduction to Nuclear and Subnuclear Physics*, W.A. Benjamin, Inc., New York, 1965.

Seife, C., *Zero - The Biography of a Dangerous Idea*, Viking Penquin, New York, 2000.

Series, G.W., *Advances - The Spectrum of Atomic Hydrogen*, World Scientific, New Jersey, 1988.

Serway, R.A., *Physics for Scientist & Engineers with Modern Physics*, 3rd Edition, Saunders Golden Sunburst Series, Saunders College Publishing, Philadelphia, PA, 1990.

Sharlin, H.I., *Lord Kelvin - The Dynamic Victorian*, The Pennsylvania State University Press, PA, 1979.

Siegfried, T., *The Bit and the Pendulum*, John Wiley & Sons, Inc., New York, 2000.

Simhony, M., *Matter, Space, Radiation*, World Scientific, New Jersey, London, 1994.

Smolin, L., *The Trouble With Physics: The Rise of String Theory, The Fall of a Science, and What Comes Next*, Houghton Mifflin, 2006.

Sprott, J.C., *Strange Attractions - Creating Patterns in Chaos*, M&T Books, New York, 1993.

Stenger, V.J., *God and the Atom – From Democritus to the Higgs Boson: The Story of a Triumphant Idea*, Prometheus Books, New York, 2013.

Sternglass, E.J., *Before the Big Bang - The Origins of the Universe*, Four Walls Eight Windows, New York, NY, 1997.

Strogatz, S., *Sync - The Emerging Science of Spontaneous Order*, Hyperion, New York, 2003.

Sunden, O., "Time-Space-Oscillation: Hidden Mechanism Behind Physics," *Galilean Electrodynamics*, Vol. 12, Special Issue 2, Fall 2001.

Swedenborg, E., *The Principia*, Swedenborg Society, London, 1912.

Synge, J.L., "The Electrodynamic Double Helix," In *Magic Without Magic: John Archibald Wheeler* - A Collection of Essays in Honor of His Sixtieth Birthday, edited by John R. Klauder, W. H. and Company, San Francisco, 1972.

T

Talbot, M., *The Holographic Universe*, Harper Perennial, 1992.

Thomson, J.J., *A Treatise on the Motion of Vortex Rings*, MacMillan and Co., London, 1883.

Thomson, J.J., *Electricity and Matter*, Charles Scribner's Sons, New York, 1904.

Thomson, D.W. and Bourassa, J.D., *Secrets of Aether*, Published by The Aenor Trust, Salem, OR, 2004.

Time-Life Books, *A Soaring Spirit - Time Frame BC 600-400*, The Time Inc.Book Company, Alexandria, VA, 1987.

Time-Life Books, *Empires Ascendant - Time Frame 600 BC - AD 200*, The Time Inc. Book Company, Alexandria, VA, 1987.

Tricker, R.A., *The Contributions of Faraday and Maxwell to Electrical Science*, Pergamon Press Ltd., London, UK, New York, 1987.

Thorne, K.S., *Black Holes & Time Warps - Einstein's Outrageous Legacy*, W.W. Norton & Company, New York, 1994.

Treasures of Early Irish Art: 1500 B.C. to 1500 A.D., From the Collections of the National Museum of Ireland, Royal Irish Academy, Trinity College, Dublin, 1977.

V

Valens, E.G., *The Attractive Universe: Gravity and Shape of Space*, Motion, Magnet, 1969.

Van Eenwik, J.R., *Archetypes & Strange Attractors - The Chaotic World of Symbols*, Inner City Books, Toronto, Canada, 1997.

Van Flandern, T., *Dark Matter, Missing Planets & New Comets - Paradoxes Resolved, Origins Illuminated*, North Atlantic Books, Berkeley, CA, 1993.

Valone, T., "Inside Zero Point Energy," *Journal of New Energy*, Vol. 5, No. 4, 2001.

Van Nostrand's Scientific Encyclopedia, D. Van Nostrand Company, Inc., 1958.

Veltman, M., *Facts and Mysteries in Elementary Particle Physics*, World Scientific, New Jersey, 2003.

Venable, W.M., *The Interpretation of Spectra*, Reinhold Publishing Corporation, New York, 1948.

Volk, G., "Toroids, Vortices, Knots, Topology and Quanta," Proceedings of of the 18th Annual Conference the NPA, 6-9 July, 2011, at the University Maryland College Park, MD, Vol. 8, 2011.

Vrooman, J.R., *Rene Descartes - A Biography*, G.P. Putnam's Sons, New York, 1970.

W

Wagner, O.E., "Structure in the Vacuum," *Frontier Perspectives*, Vol. 10, No. 2, Fall 2001.

Walker, F.L., "The Fluid Space Vortex: Universal Prime Mover," *Physics Essays*, Vol. 15, No. 2, 2002.

Watson, J.D., *The Double Helix*, W.W. Norton & Company, New York, 1980.

Weinberg, S., *Dreams of a Final Theory*, Pantheon Books, New York, 1992.

Westfall, R., *The Life of Isaac Newton*, Cambridge University Press, New York, NY, 1994.

Wheeler, J.A., *Geometrodynamics, Topics of Modern Physics, Vol.1*, Academic Press Inc., New York, NY, 1962.

White, H.E., *Introduction to Atomic Spectra*, McGraw-Hill Book Company, Inc., New York, 1934.

Whitney, S.K., "9 Editor's Essays," *Galilean Electrodynamics*, Vol. 16, Special Issue 3, Winter 2005.

Whitney, S.K., *Algebraic Chemistry – Applications and Origins,* Nova Science Publishers, Inc., New York, 2013.

Wiener, N., *Cybernetics or Control and Communication in the Animal and the*

Machine, 2nd edition, The MIT Press and John Wiley & Sons, Inc., New York, 1961.

Woit, P., *Not Even Wrong: The Failure of String Theory and the Search for Unity in Physical Law*, Basic Books, 2006.

Wolff, M., *Exploring the Physics of the Unknown Universe*, Technotran Press, Manhattan Beach, CA, 1990.

Wolff, M., "Origin of the Mysterious Instantaneous Transmission of Events in Science," *The Cosmic Light*, Vol. 4, No. 2, Spring 2002.

Wolfram, S., *A New Kind of Science*, Wolfram Media, LLC, Champaign, IL, 2002.

Z

Zeman, R.K. et al., *Helical/Spiral CT - A Practical Approach*, McGraw-Hill, Inc., New York, 1995.

Zombeck, M.V., *Handbook of Space Astronomy and Astrophysics*, Cambridge University Press, Cambridge, UK, 1990.

Zwikker, C., The Advanced Geometry of Plane Curves and Their Applications, Dover Publications, Inc., New York, 1994.

www.ingramcontent.com/pod-product-compliance
Lightning Source LLC
Chambersburg PA
CBHW022054210326
41519CB00054B/363